ESSAYS

ESSAYS

Papers and Presentations Exploring the Vedic Tradition and Modern Science
(1977–2007)

Richard L. Thompson, Ph.D.
(Sadāpūta dāsa)

RLTA PRESS
Alachua, Florida

The publishers of this volume would like to offer their profound appreciation to the Bhaktivedanta Book Trust trustees for their dedicated support of the Richard L. Thompson Archives and the Bhaktivedanta Institute for Higher Studies, which has facilitated this project.

A special thanks to Guru Das, for his encouragement and support.

For more information about the Richard L. Thompson Archives please contact:
info@richardlthompson.com • www.richardlthompson.com

For more information about the Bhaktivedanta Institute please contact:
info@bihstudies.org • www.bihstudies.org

On the Cover: Illustration by Richard L. Thompson is in Chapter 8, page 136. The author describes the image as a "Poetic description of Jambūdvīpa in the Śrīmad-Bhāgavatam as a lotus flower." Rear cover illustrations are located on pages 154, 160, and 162.

ISBN 978-1-959829-07-2

Cover and book design:
Eight Eyes
www.eighteyes.com

Dedicated to

His Divine Grace
A. C. Bhaktivedanta Swami Prabhupāda

oṁ ajñāna-timirāndhasya jñānāñjana-śalākayā
cakṣur unmīlitaṁ yena tasmai śrī-gurave namaḥ

Books by Richard L. Thompson

Consciousness: the Missing Link (1980)
by His Divine Grace A. C. Bhaktivedanta Swami Prabhupāda,
Dr. T. D. Singh, and Richard L. Thompson

Mechanistic and Nonmechanistic Science:
An Investigation into the Nature of Consciousness and Form (1981)

Vedic Cosmography and Astronomy (1989)

Parallels [Alien Identities]:
Ancient Insights into Modern UFO Phenomena (1993)

Forbidden Archeology (1993)
by Michael A. Cremo and Richard L. Thompson

Mysteries of the Sacred Universe:
The Cosmology of the Bhāgavata Purāṇa (2000)

Maya:
The World as Virtual Reality (2003)

God And Science:
Divine Causation and the Laws of Nature (2004)

[A]s demonstrated by Dr. Richard Thompson of the State University of New York at Binghamton and confirmed by several Nobel laureates in physics who have praised his work, the laws of nature governing the transformation of matter simply do not contain sufficiently complex information to account for the inconceivable complexity of events taking place within our own bodies and those of other life forms. In other words, not only do the material laws of nature fail to account for the existence of consciousness, but they cannot explain even the interaction of material elements at complex organic levels.

– *Śrīmad-Bhāgavatam*, Canto 12, Verse 43 purport
The Bhaktivedanta Book Trust, Los Angeles (1988)

Contents

Introduction

Dr. Richard L. Thompson (1947–2008), also known by his Vaiṣṇava name Sadāpūta dāsa, was a mathematician, scientist, philosopher, author, researcher of ancient cosmology, and devoted practitioner of *bhakti-yoga*. In 1974 he received his PhD specializing in probability theory and statistical mechanics from Cornell University. During this time, he found inspiration in the philosophy of *Bhagavad-gītā* and became a dedicated follower of the Vaiṣṇava tradition. He later served as a founding member of the Bhaktivedanta Institute established by A. C. Bhaktivedanta Swami Prabhupāda, Founder-*Ācārya* of the International Society for Krishna Consciousness (ISKCON), more popularly known as the Hare Krishna Movement.

Materials in the Richard L. Thompson Archives (RLTA), held in Alachua, Florida (see footnote 1), offer valuable insight into Thompson's research interests in the discourse engaging religion, science, and the natural world. The Archives include articles published in both well-established peer-reviewed scientific journals as well as scholarly journals that might be considered avant-garde. Its most prominent feature are his extensive writings exploring the theistic tradition of Gauḍīya Vaiṣṇavism, focusing on the long discourse between science and religion with specific reference to the Vaiṣṇava tradition. For a full biographical essay on the life of Richard L. Thompson, please see: www.richardlthompson.com/full-biographical-essay, hosted on the RLTA website.

Thompson kept extensive files of his prolific output that were carefully held by his family, who subsequently presented them to the Archives to preserve, organize, and digitize for public access, most notably on the RLTA website, www.richardlthompson.com. RLTA projects include cataloging his papers and other research materials, an oral history project, the maintenance and continued development of the archival website, and fostering ongoing research with reference to his legacy as a founding member of the Bhaktivedanta Institute. Thompson's work has not only influenced important ISKCON projects such as the Temple of the Vedic Planetarium (see tovp.org), but offers potential to inspire future research.

The initial chapters included in Part I of this volume offer a selection of

Thompson's work on behalf of the Bhaktivedanta Institute. Some of it was featured in its own literature and seminars as well as in scholarly conferences and journals that engaged a cross-disciplinary exploration of the cultural, philosophical, and intellectual traditions of the Indian subcontinent. Many of these papers might have received less exposure compared to other works, having not been previously included as chapters in his formal publications or featured as articles in ISKCON's *Back to Godhead* magazine.

The first essay in this section includes Thompson's introductory and concluding chapters from the 1977 Bhaktivedanta Institute Monograph Series Number 3 with the title, "Consciousness and the Laws of Nature." It was written for the Institute's inaugural "Life Comes from Life" conference hosted at ISKCON's Krishna Balaram Mandir in Vrindavan, India, that year. This monograph series represents the Institute's earliest example of disciplined scholarly outreach.

The second essay, "The Nature of Biological Form," was the lead article for the January 1979 Bhaktivedanta Institute Bulletin, Vol I. No. 1. In it, Thompson considered how Darwin did not offer a comprehensive analysis of the intermediate, transitional forms for the development of various complex organs. Thompson proposed basic complexity arguments to help illustrate the difficulties in forming sophisticated biological structures invoked by evolutionary theory. He suggested that texts like the *Bhagavad-gītā* can offer "a unified description ... of an agency that accounts for the origin" of complexity in "both biological form and human agency."

The third selection, "Numerical Analysis and Theoretical Modelling of Causal Effects of Conscious Intention" (1991), reviewed interactions of human consciousness with sensitive physical devices as reported by Robert Jahn and Brenda Dunne at the Princeton Engineering Anomalies Research (PEAR) program managed under the aegis of Princeton University's School of Engineering and Applied Science. Although parapsychological theories considered by Jahn did not gain main-stream acceptance, Thompson applied techniques drawn from statistical mechanics that helped to explain some of the phenomena discussed in these reports. His analysis supported the hypothesis that events such as these may well suggest a number of possibilities that could potentially account for these anomalies.

In the fourth chapter, Thompson ventures into the field of archeology in his essay, "Emperor Aśoka and the Five Greek Kings" (1994). While the Archives has been only able to locate a nearly finished draft of this paper, it none-the-less offers unique perspectives concerning the Aśokan royal edicts that were carved on rocks and pillars throughout the Indian subcontinent. While drawing upon contemporary analyses of Indian history, Thompson proposed that the names for five kings noted in these inscriptions, generally identified with Hellenistic kings from the third century BCE, might also refer to a far more ancient lineage.

Chapter 5, "On the Antiquity of Star Coordinates from Indian *Jyotiṣa Śāstras*," was presented at the 1994 International Symposium of Ancient Indian Chronology held at the B. M. Birla Science Centre in Hyderabad, India. It explored a comparison between star coordinates listed in traditional Indian astronomical texts (*Jyotiṣa Śāstras*) and the coordinates of corresponding stars listed in modern tables. When applying proper motion of the stars and precession of the equinox, modern star positions show a tendency to move towards the *jyotiṣa* star positions as we go back in time. Thompson argued that this may suggest that the *jyotiṣa* coordinates were measured in the distant past, well prior to what would normally be appreciated as recorded human history. The time period from 25,000 to 55,000 years ago appears compatible with the adjusted star coordinates.

"Planetary Diameters in the *Sūrya-siddhānta*" (this version published in 1997) features Thompson's analysis of mathematical correspondences between modern astronomical knowledge and similar information potentially embedded in ancient texts. Thompson reported that by mathematically combining the circumferences of the planetary orbits along with the angular planetary diameters, both of which are given in specific verses of the text, he was able to calculate diameters for the planets that correspond well with modern figures. Thompson hypothesized that the angular diameter rule presented in the *Sūrya-siddhānta* may have been based on sophisticated astronomical knowledge that had been subsequently lost until more recent times.

The seventh paper, "Anomalous Textual Artifacts in Archeo-Astronomy," was presented at the 1996 World Association of Vedic Studies conference. The theme for the conference, "Revisiting Indus-Sarasvati Age and Ancient India," fits well with Thompson's discussions about how archeological evidence can survive within ancient texts. Examples of potential textual artifacts include the analysis offered in his "Planetary Diameters" paper presented in Chapter 6, along with a new study of mathematical similarities between the observable orbits of the planets from a geocentric perspective and the dimensions of the rings of Bhū-maṇḍala described in the *Bhāgavata Purāṇa*. Thompson proposed that this evidence can support arguments for the existence of sophisticated astronomical knowledge in ancient times.

PART II: PAPERS AND PRESENTATIONS offers a collection of three seminar presentations along with two unpublished papers.

The first, in Chapter 8, "Points Concerning the Vedic Conception of the

Universe" (c. 1984), is a facsimile reprint of a hand-written, forty-one-page document that includes fifty-seven hand-drawn illustrations by the author. This early sketch appears to provide a roadmap for Thompson's future research and also anticipates many of the research developments during the past forty years by scholars interested in these topics. Around this time, Thompson would also describe his motive for undertaking this type of research as an attempt "to resolve the apparent contradictions between these two sources of Vedic astronomical knowledge [*Śrīmad-Bhāgavatam* and *Sūrya-siddhānta*] and to arrive at a clear understanding of the Vedic conception of the universe."[1] A typed version of the document is offered side-by-side opposite each page of the facsimile reprint of the hand-written manuscript.

Chapter 9, "On the Relationship Between Consciousness and Matter" was presented at a 1985 meeting of guests and science faculty at the Eindhoven University of Technology in the Netherlands. There, Thompson discussed the phenomenon of consciousness within the framework of a mechanistic paradigm. After considering arguments offered by Walter Elsasser, Eugene Wigner, and David Bohm, Thompson introduced perspectives from the Vedic tradition, in particular that of the *jīva* and Paramātmā, as part of a potential explanatory paradigm. An engaging discussion followed the talk, which explored a variety of research topics that could be considered while utilizing this approach.

The next section, "A Trans-temporal Approach to Free Will and the Laws of Physics" (10a), contains a presentation at the 1990 "Consciousness within Science" conference hosted by the Bhaktivedanta Institute in San Francisco. This gathering featured the participation of over a dozen eminent scientists and scholars.F Thompson's talk addressed ongoing contentions involving the sense free agency and the restrictions imposed by natural law. He proposed a dualistic approach influenced by concepts drawn from deterministic chaos theory, which he felt could help illustrate how the effect of consciousness on matter could apparently violate the laws of physics without violating the conservation of energy principle. After introducing the mathematical application known as global non-linear optimization, he suggested that "will" can act in a fashion similar to a sub-atomic constraint, which then would affect the workings of nature.

Section (10b) offers a facsimile of an unpublished technical paper that provides more rigorous mathematical support for the presentation in (10a). In it,

1 Richard L. Thompson, "The Opinions of Previous Vaisnava Acaryas Concerning Vedic Cosmology and Astronomy," Box 1 folder 9, Richard L. Thompson Archives, 12834 NW 151 Rd., Alachua FL 32615.

Thompson presented a basic formulation of classical and quantum mechanics in which these two theories take on a similar form. Thompson's Archives also contain two earlier versions of this analysis dated from 1986. A less technical lecture, "God and the Laws of Physics," would be given at the 1986 First World Congress on the Synthesis of Science and Religion held in Bombay, India. Thompson published a reprint from the 1986 conference proceedings as Chapter 1 in his compilation volume, *God & Science: Divine Causation and the Laws of* Nature (2004). There he commented, "The technical details of this formulation of quantum mechanics are presented in an unpublished paper. Formally, it adds nothing to quantum mechanics, but it shows how this theory can be seen to provide a nondeterministic account of objective reality.[2]

Thompson's final presentation, "Interpretation and the *Śrīmad-Bhāgavatam*," was given at the "Second Annual Conference of the ISKCON Academy of Arts and Sciences," held in West Virginia in December 2007. There, Thompson addressed concerns involving interpretation of ancient texts, particularly when analyzing evidence presented in the *Śrīmad-Bhāgavatam* that can appear at odds with professional scientific methodologies. He then discussed several ways to explore these challenges by using examples from Purāṇic literature such as its description of "calculations of time from the atom." Thompson argued that by maintaining a mature flexibility while respecting the integrity of each tradition, scholars can consider these seemingly disparate perspectives in a manner that potentially fosters an enhanced picture of both.

Although Richard Thompson's reputation both as a scientist, and as an advocate for *bhakti-yoga*, have been well-established by his numerous publications that include over six books, four monographs, thirty professional papers, and nearly forty *Back to Godhead* articles, the essays included in this volume can also add insight and potentially fresh perspectives on the discourse between science and religion with specific reference to the Vaiṣṇava tradition.

2 Richard L. Thompson, *God & Science: Divine Causation and the Laws of Nature* (Alachua, FL: Govardhan Hill Publishing, 2004), 16.

PART I

ESSAYS

CHAPTER 1

Consciousness and the Laws of Nature

The Bhaktivedanta Institute Monograph Series Number 3

Chapter I. "Introduction"
and VII. "The Laws of Consciousness and Matter"

Published by:
Bhaktivedanta Institute
Boston • Bombay (1977)

Notes:

This monograph was recently published in a compilation edition titled, *The Bhaktivedanta Institute Monograph Series* (see richardlthompson.com/book/bhaktivedanta-institute-monograph-series).

Due to space considerations, only Chapters I. "Introduction" and VII. "The Laws of Consciousness and Matter" have been included in this volume. The full monograph is available on the Richard L. Thompson Archives website: richardlthompson.com/1977-consciousness-and-laws-nature-bhaktivedanta-institute-monograph-series-number-3.

I.

Introduction

> It is premature to reduce the vital process to the quite insufficiently developed conception of 19th and even 20th century chemistry and physics.
>
> —*Louis de Broglie*

At the present time it is widely claimed that life can be understood simply as a complicated interaction of atoms and molecules in accordance with known physical laws. High school and college textbooks of biology begin with the study of chemical bonding, proceed on to molecular evolution, and flatly assert that scientists have "been able to synthesize the stuff of life in a laboratory flask."[1] In scientific books and journals the theory that life is a combination of material elements is widely accepted as nearly unquestionable fact. The biochemist James Watson sums up this viewpoint as follows:

> We see not only that the laws of chemistry are sufficient for understanding protein structure, but also that they are consistent with all known hereditary phenomena.
>
> Complete certainty now exists among essentially all biochemists that the other characteristics of living organisms (for example, ... the hearing and memory processes) will all be completely understood in terms of the coordinative interactions of small and large molecules.[2]

Watson later goes on to assert that these molecular interactions have been understood by the modern theory of quantum mechanics.

> The various empirical laws about how chemical bonds are formed were put on a firm theoretical basis. It was realized that all chemical bonds, weak as well as strong, were based on electrostatic forces.[3]

In this paper we argue that this view of life is extremely shortsighted. Not only has life not been understood as a product of matter, but our understanding of

1 Sherman, *Biology: A Human Approach,* p. 4.
2 Watson, *Molecular Biology of the Gene,* p. 67.
3 Watson, p. 105.

matter itself is seriously deficient. We will show, in fact, that in order to remedy the deficiencies in our concept of matter we are forced to adopt an understanding of life completely different from the accepted scientific view.

We will review some of the important theories of nature of the modern scientific age. This review will culminate in a more detailed account of the present dominant theory of quantum mechanics. We shall see that none of these theories have been successful. The theories preceding quantum mechanics have all been rejected for various reasons, and the quantum theory itself possesses serious defects which rule it out as a valid understanding of nature.

Since the time of Newton all major scientific theories of nature have been characterized by two assumptions:

(1) All of the significant features of nature can be described by numbers.
(2) All of the phenomena of nature are governed by laws which can be described by very simple mathematical equations relating these numbers to one another.

Furthermore, throughout the history of modern science, scientists have strongly tended to assume that all phenomena can be accounted for (at least in principle) by the accepted laws of their day.

These assumptions form the foundation for the modern scientific view of the absolute truth. The *absolute truth* can be defined as the ultimate causative principle or agency underlying all of the phenomena of nature; and the understanding of this fundamental cause can be seen as the goal of all fundamental research in science. However, conditions (1) and (2) impose a very severe *a priori* restriction on the nature of the absolute truth. There is no particular reason to suppose that every significant feature of nature can be described by numbers, or that those which can be so described are governed by simple equations. Our thesis is that nature cannot actually be understood within the framework imposed by these conditions.

In particular, we are proposing that the phenomenon of consciousness cannot be described by numbers, and that the behavior of matter is less and less amenable to description by simple equations the more intimately it is associated with consciousness. Since consciousness is a feature of life (at least on the human level) this thesis directly contradicts the theory that life is only a product of chemical reactions obeying simple physical laws. It is perhaps ironic, then, that compelling support for it is to be found not in the science of biology (in this paper, at least) but in physics, the fundamental study of inanimate matter.

As we shall show, consciousness is directly involved with basic problems in the quantum theory which cannot be resolved within the framework of conditions (1) and (2). Since this theory is solidly based on these assumptions, it cannot be correct as it stands. As a solution to these problems we will therefore outline a description of nature in which consciousness appears as a primary, irreducible feature of the absolute truth, lying beyond the reach of mathematical description.

In this description, consciousness enters as a natural analogue of the basic laws of nature figuring in the conventional theories of physics. It thus plays the role of an active constituent of nature. The following chart compares the role of consciousness in this view to the principle of electrical interaction in standard physics.

Entities	Electrons	"Quanta" of Consciousness
Principles of interaction	Electric field	Absolute consciousness

Just as the electrons interact with other matter through the agency of the electric field (in the standard theory), so the individual conscious entities, or "quanta" of consciousness, interact with matter through the agency of absolute consciousness. However, whereas the electrical interaction is described by certain simple equations (called Maxwell's equations in the standard theory), the interaction of the conscious entities cannot be described in this way. This mathematical indescribability will be reflected by the behavior of matter—insofar as matter is affected by these interactions, its behavior will also defy reduction to any simple mathematical scheme.

Although this description of nature forms a natural extension of the standard theories, it is also consistent with a much older conception of nature epitomized by the ancient Sanskrit text, the *Bhagavad-gītā*.[4] As such it has many profound implications about the nature and potentialities of life which take us far beyond the limited schemes of modern biological theory. We feel that it deserves serious consideration both for this reason and for the elegant and natural way in which it resolves basic difficulties in modern physics.

Once when the physicist Niels Bohr was asked whether the known laws of physics would account for life, or whether life involved some higher principles as yet unknown, he replied that he did not know. He went on to say, however, that a scientist must be very conservative in his thinking, and very hesitant to discard old concepts or adopt new ones unless compelled to do so by overwhelming

4 A. C. Bhaktivedanta Swami Prabhupada, *Bhagavad-gītā As It Is.*

evidence.[5] He said that we should therefore act on the assumption that the known laws of physics, as embodied in the theory of quantum mechanics, would suffice to give us a complete understanding of the phenomena of life.

We would like to suggest that Bohr's conservatism was misplaced. Bohr was proposing that conservatism means sticking to the most radical and speculative theory that science had yet produced—a theory which was poorly understood, full of unresolved paradoxes, and tested only in limited circumstances.

A much more fruitful conservative approach entails the ancient understanding that the fundamental principle of life is an entity—the self or *atma* (quantum of consciousness)—which is not reducible to matter. As such, this approach is not merely a theoretical exercise, but it has many practical, empirical consequences. In particular, it entails the direct observation and study of the *atma* and its relationship with the absolute consciousness, or *paramatma.* Some of the principles governing this study are briefly outlined in the last section.

Bibliography

(Introduction)

A. C. Bhaktivedanta Swami Prabhupada. *Bhagavad-gītā As It Is.* New York: Collier Books, 1972.

Heisenberg, W. *Physics and Beyond.* New York: Harper and Row, 1971.

Sherman, I. W. & V. G. *Biology—A Human Approach.* New York: Oxford Univ. Press, 1975.

Watson, J. D. *Molecular Biology of the Gene,* 2nd ed. Menlo Park, Calif.: W. A. Benjamin, Inc., 1970.

5 Heisenberg, *Physics and Beyond,* p. 112.

VII.

The Laws of Consciousness and Matter

> The absolute controller is acting within the heart, and is directing the wanderings of all living entities, who are seated as on a machine, made of the material energy.
>
> — *Bhagavad-gītā*

In this review of various scientific theories, we have been led to the conclusion that there are fundamental defects and limitations in man's present scientific understanding of matter and its laws. As we have seen, many theories have been proposed and rejected in the past, and this process still goes on. We have seen that the present dominant theory of quantum mechanics suffers from serious defects in its fundamental structure which can only be partially remedied by artificial, stop-gap measures. Its claim to universal applicability is unconfirmed, and there exists evidence which directly contradicts this claim. We can thus conclude that scientists do not know what matter is nor how it is acting.

The basic philosophical presupposition of modern science is that all the effects of nature are the consequences of a few simple laws capable of mathematical expression. Our thesis is that this presupposition has by no means been established. Indeed, as stated by the physicist D. Bohm in a discussion of this point, "the historical development of physics has not confirmed the basic assumptions of this philosophy, but rather has continually contradicted them."[1] In this section we would like to explore some of the possibilities that arise if we discard this philosophical assumption and suppose instead that there may be no limit to the variety of natural laws and entities. In particular, we would like to consider the role of consciousness as a phenomenon of nature.

The Irreducible Character of Consciousness

Throughout its struggles with the nature of matter, modern science has neglected consciousness almost completely, even though this phenomenon is the most primary feature of our existence as living beings. Indeed, the very existence of consciousness has proven to be a great embarrassment to the theoreticians of quantum mechanics. Some physicists, such as Niels Bohr, have been content

---

1 Bohm, p. 131.

to ignore, or "renounce," the very question of understanding consciousness. Others, such as von Neumann and Wigner, have recognized that consciousness lies outside the domain of their theories, but must be taken into account if a true understanding of nature is to be reached. However, they have not been able to introduce consciousness into their theoretical picture in a satisfactory way.

We would like to suggest that consciousness is a feature of reality which is incapable in principle of being adequately described in numerical terms. In our review of different mathematical descriptions of nature we have dealt with essentially two types of theories:

(1) In the first type there is a one to one correspondence between certain sets of numbers and the "fundamental elements" of reality. In such a theory every existing feature of nature can in principle be mathematically represented, for every feature must be a combination of fundamental elements. The classical nineteenth century theories were generally of this type.

(2) In the second type there is a correlation, which may be statistical, between certain numbers in the theory and the quantitative results of experimental measurements. It is not possible, however, to pin down the underlying reality which gives rise to these results. The calculations of the theory are simply supposed to predict the results to the greatest possible extent, and their structure cannot be thought of in terms of underlying entities or phenomena. The Ptolemaic astronomical system is an example of this kind of theory.

Generally, theories are interpreted as being of the first type. However, many people have felt impelled to regard quantum mechanics as a theory of the second category. (This is the option of admitting the wave function to be an incomplete description of nature, and simply accepting the whole theory as a set of approximate calculations.) Maxwell's electromagnetic theory can also be viewed in this way, as we pointed out in section II.

A theory of the second type may predict bodily movements or the electrical potentials of neurons in the brain very accurately (although we should stress that no such predictions have actually been made from any existing theory). However, it can give us no insight at all about how or why conscious awareness is associated with such phenomena. From a calculated list of numbers corresponding to some physical behavior, what can we say about the awareness that may or may not have been associated with that behavior? It remains a complete mystery.

In a theory of the first type we are confronted with a picture of the world as a composite of many simple, elemental entities. As we have noted before, there is no reason to suppose that conscious awareness will exist just because many of these entities are juxtaposed in a certain pattern. Each simple, thoroughly insentient entity interacts with the others by some simple mechanical rule. At any one time each entity is changing (in position, orientation, amplitude, spin, or whatever) in a simple, thoroughly insentient way depending on this rule. No entity "knows" in any sense what the others are doing. How then can clear conscious awareness exist as a consequence of the presence of many such entities in some spatial arrangement?

The essential motivating idea for supposing that consciousness can be "explained" in this way is the notion that conscious awareness somehow corresponds with physical behavior. We have touched on this point before. However, to further dispel this misleading conception, let us consider a particular form in which it often appears. This is the idea that a computer can be conscious if it simulates by calculation the appropriate physical events occurring in a person's brain.

The British mathematician A. M. Turing has advanced the argument that all of a person's behavior can be duplicated by a suitably programmed computer.[2] Of course, this is far from being actually demonstrated. But, if we were confronted with such a computer, which could talk with us and exhibit all of the symptoms of human personality, then we might indeed be tempted to suppose that it possessed human consciousness. The question is: would this supposition be justified?

In order to answer this, consider what is going on within the computer during its calculations. Within the computer's "memory" unit there is stored a list of numbers encoding instructions for simple logical and arithmetical operations. Part of this list might look as follows:

code number	meaning
10 4787 0648	"Add the number at location 4787 to the number at 0648."
02 0648 1246	"If the number at 0648 is positive go to the step stored at 1246; otherwise, go to the next step."
03 0648 1267	"If the number at 0648 is negative go to the step stored at 1267; ..."
	... etc... .

2 Turing, "Computing Machinery and Intelligence."

All that the computer is doing at any one time is mechanically (or electrically) carrying out the instruction corresponding to one of these code numbers. The total behavior of the computer is simply the net result of the execution of many of these instructions, one after the other.

Since only a few simple electrical interactions are taking place at one time, it is hard to see how the computer could be conscious. If the computer were slowed down (as is possible) so that each simple step was stretched out over several seconds, the pattern and sequence of the steps would remain the same. Since the behavioral output of the computer would be slower but otherwise the same, does it follow that the conscious awareness of the computer would simply be stretched out in time? If not, we would have to explain why executing the instructions at one speed would generate conscious awareness of the thoughts being simulated, while at another speed there would be no consciousness of these thoughts.

Also, changing the construction of the computer should presumably not affect its consciousness as long as it is programmed to carry out the same steps, for this assures that its behavior will exhibit the same pattern. Figure 28 illustrates one form in which a computer can be constructed. Here the computer instructions are used to set up a gigantic "game" which could be played step by step by a child. As the child carries out these steps, will the same consciousness of the simulated thoughts be manifested there—stretched out, perhaps, over several years? This hardly seems plausible, but otherwise how are we to judge which of many computers with equivalent programs will be conscious and which ones will not?

We would like to suggest then that consciousness must be due to some existing entity in nature that cannot be numerically described. In the remainder of this paper we would therefore like to explore the implications of the following assumption: let us suppose that there are primary, irreducible entities that possess conscious awareness. These can be thought of as *quanta of consciousness* in analogy to electrons which, in standard physical theory, can be thought of as irreducible quanta of electricity. Each of these quanta of consciousness carries or possesses the individual awareness of a particular individual living being. We shall also use the term *ātma* to designate one of these conscious entities.

Although we are assuming the conscious entity to be primary and irreducible, it is not simple like an electron. Whereas the electron is attributed simple properties, such as charge and spin, the quantum of consciousness must be capable of comprehending and appreciating very complex situations. For this reason we shall designate the quanta of consciousness collectively as the *superior energy*, in contrast with matter, which can be called the *inferior energy*.

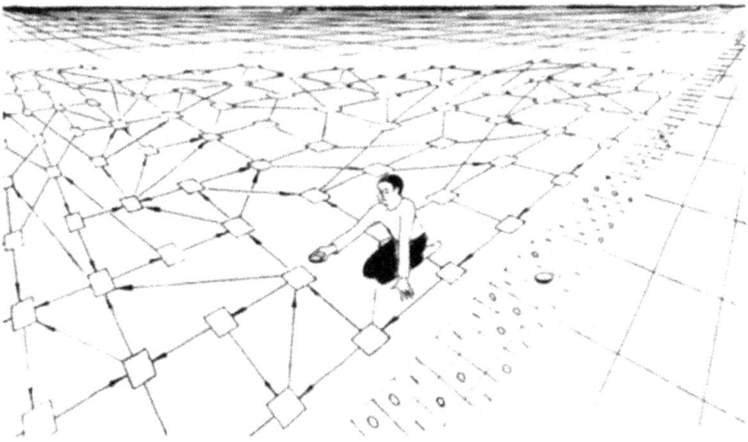

Figure 28. A computer in the form of board game. Suppose that the program, printed in steps on the squares of the board, is intended to recreate the consciousness of Joe Smith. As the game is played, will his consciousness be present there?

Our proposal, then, is that each conscious living organism consists of a body composed of matter plus an *ātma* which associates with this body and is conscious of the activities of the bodily senses. Many conscious entities must be invoked to account for the observed existence of many individual conscious beings. The content of the consciousness of a given *ātma* interacting with matter will depend on the physical arrangement of that matter. For example, there will be consciousness of bodily sense perception if the senses are in working order, but there will not be such consciousness if they are not working. This state of affairs can be called *materially conditioned consciousness.*

By "matter" we mean the familiar object of study of physics and chemistry. Matter can be conceived of as an inherently insentient type of energy that can be transformed into many different temporary forms and configurations. The theory of quantum mechanics has left our understanding of matter rather "fuzzy" to say the least. It is capable to some extent of being described by various mathematical laws. However, it is evident that matter is still in many respects a mystery to modern science.

We should note that many of the features of our experience that are commonly called "mental" may be part of the materially determined content of consciousness, rather than part of the conscious entity, or *ātma* , itself. This would include many different temporary features of our mental life such as the memory of words, and different skills and habits. However, the *ātma* must be inherently capable of awareness and appreciation of these things, or they would simply go unnoticed.

Our basic proposition is that the quantum of consciousness cannot be described mathematically as a combination of insentient entities. The question naturally arises of whether or not the *ātma* might be a combination of insentient, mathematically indescribable entities which fit together in a mathematically indescribable fashion. Actually, this is a rather cumbersome and intractable proposal. Since consciousness exists but cannot be numerically described, it follows that some kind of conscious entity must exist which cannot be understood in our familiar mathematical terms as a combination of simple, insentient entities. The simplest and most economical solution to this problem is to suppose that consciousness is an absolute feature of reality, and that the conscious entities are completely irreducible. This is much more in accord with Occam's razor, the principle of economy of thought, than is the introduction of many other completely mysterious irreducible entities.

Consider, for example, the hypothesis that consciousness is an "epiphenomenon" of brain activity.[3] This hypothesis is outlined in Figure 29. In this figure the laws governing matter are taken, for the sake of argument, to be of the classical 19th century type. The point of the epiphenomenon hypothesis is that certain material configurations, such as functioning human brains, "generate" consciousness, while others do not. The consciousness, on the other hand, is presumed to have no effect on matter (and therefore is not included among the arguments of the laws of matter, F1 and F2.)

Laws of matter:

$$\frac{\partial p}{\partial t}j = F_1\,(p_1, \ldots, p_n; q_1, \ldots, q_n; j)$$

$$\frac{\partial q}{\partial t}j = F_2\,(p_1, \ldots, p_n; q_1, \ldots, q_n; j)$$

Law generating consciousness:

$$G\,(p_1, \ldots, p_n; q_1, \ldots, q_n) = \begin{cases} \text{"consciousness"} \\ \text{"unconsciousness"} \end{cases}$$

Figure 29. A model in which consciousness figures as an "epiphenomenon."

In Figure 29, the law, G, which generates consciousness in this way is indicated. If G is a fundamental natural law, then G must eternally exist in some sense. Yet G must also be mathematically indescribable since it generates consciousness, which is mathematically indescribable. G must somehow generate individualized, mathematically irreducible conscious awareness in association with each suitable brain. It is very hard to obtain a clear understanding of this mysterious G, or relate it, for example, to the theory that life has originated from matter by evolution—one might wonder what G was doing during all the time when no brains existed. The assumption of irreducible conscious entities certainly poses no more difficulties than the hypothesis of such a G.

However, this assumption has the interesting implication that the conscious self survives death. In fact, if we suppose that the *ātma* tends to be associated

3 See *The Encyclopedia of Philosophy,* vol. 5, p. 343.

with highly organized material bodies—and this certainly seems reasonable from an empirical standpoint—then the transmigration of the conscious self through a succession of bodies is implied. Even though these are very striking implications, we propose to adopt the assumption in this paper that the quantum of consciousness is completely irreducible. As we shall see, this assumption dovetails with a very concise and elegant world picture that is full of significant consequences and interesting avenues of further study.

The properties of the *ātma* can be summarized as follows:

(1) The elemental carrier of consciousness.
(2) Exists in unlimited numbers.
(3) Cannot be created or destroyed (conservation principle).
(4) Tends to be associated with very complex bodies composed of material elements.

Let us consider what sort of laws of interaction are involved with the quanta of consciousness. We shall approach this from the point of view of the basic question, "What is the nature of the absolute truth, or the final cause of all causes?" Here are a number of features characteristic of the concept of the absolute truth as we have encountered it thus far in normal science:

(1) The absolute truth exists, but is inconceivable to the human mind.
(2) It is all-pervading in space.
(3) It is invariant in time.
(4) It is the source and controller of all manifestations.
(5) It possesses an inherent unity.

As we have pointed out in previous sections, the natural laws have played the role of the final, absolute cause in the various mathematical theories of physics. Certainly the laws of all basic physical theories have shared properties (1), (2), and (3). Since the natural laws have always had to be supplemented by initial conditions (or even "interim" conditions!) they have not satisfied condition (4), however. Nonetheless, scientists have generally tried to minimize the ultimate importance of the initial conditions as much as possible by means of the idea of evolution. In this way they have tried to attribute property (4) to their systems of laws to the greatest possible extent.

Likewise, the laws of physics, inasmuch as they consist of a series of apparently unrelated mathematical expressions, do not satisfy (5). However, the creators of theories have customarily tried to formulate their laws in such a way that they possess as much unity as possible. For example, the Hamiltonian formulation of Newton's laws in equations (1) and (2) of section II was considered to be a great accomplishment because of its unity and simplicity of form. Albert Einstein, in fact, devoted much of his efforts to the development of a "unified field theory" for physics because he felt that the ultimate cause underlying nature must be a harmonious unit rather than a disjointed mélange of unrelated things.

In this context, the simplest hypothesis that we could make about the absolute truth is the following: let us suppose that a "higher order law" governs the actions of both the inferior and superior energies, and that this higher order law satisfies conditions (1) through (5).

Since we are dealing with mathematically indescribable entities, this must be a higher order law of the third category mentioned in section IV. If the interaction of the *ātma* with matter proceeds according to such a law, then we might expect the behavior of matter to be more and more difficult to describe mathematically the more directly and intimately the matter is involved with the activities of the *ātma.* The more closely a mathematical approximation predicts the measurable, external manifestations of laws of this category, the more elaborate the approximation must be. Therefore, while simple laws may be applicable to inanimate matter in standard laboratory situations, we may expect that they will not suffice to describe the behavior of matter within the bodies of living organisms. This corresponds to the view of the physicist E. Wigner that "the present laws of physics are at least incomplete without a translation into terms of mental phenomena. More likely they are inaccurate, the inaccuracy increasing with the increase in the role which life plays in the phenomena considered."[4]

We have proposed that the "higher order law" we are considering should fully satisfy the criteria (1) through (5) for the absolute truth. This means that this "law" must generate and control all manifestations and also be highly unified. Certainly, unity in the sense of simple equations is not possible here, for we must expect even approximations to the material actions of this law to be highly complex, the law itself being completely beyond mathematical description. It might seem very difficult to discuss or even conceive of such an entity.

It turns out, however, that the simplest possible choice for our "higher order law" satisfies points (1) through (5) perfectly. We have already introduced consciousness as a primary, irreducible feature of reality. The most economical hypothesis for the absolute causal principal is therefore the following: let us

4 Wigner, "Physics and the Explanation of Life," p. 44.

suppose that the absolute truth is all-pervading, universal consciousness. We shall designate this universal consciousness as *paramātmā*.

Let us suppose that such things as desire, will, and purpose are inherent aspects of the conscious entity, or *ātma*. On this hypothesis, the interaction of the quanta of consciousness with matter can be understood as follows: the *ātma* situated within a particular material body exhibits certain desires; these desires are perceived by the *paramātmā*, and in coordination with the desires of all other living entities, the *paramātmā* directs the material elements of the body accordingly. This is possible since the *paramātmā* is the ultimate causative factor lying behind both the inferior and superior energies. Briefly, we are proposing that the interactive coupling between the *ātma* and matter proceeds through consciousness.

In this picture, property (5) is satisfied since the *paramātmā* is one conscious unit that perceives all things. We were already forced to posit an indivisible unit that can perceive a multiplicity of things when we introduced the *ātma*, so this is not something qualitatively new. In this picture the individual *ātmas* can be regarded as minute quantized parts of the *paramātmā* which share its properties on a small scale.

We have also posited that the *paramātmā* should satisfy property (4) of the absolute truth. By introducing laws of higher mathematical complexity—what to speak of laws transcending mathematical description—we have already been forced to depart from the spirit of evolutionary thinking. The idea of evolution is basically that complex form arises by simple processes from a situation where there was no such form. Here, however, we have already introduced absolute complexity. By proposing that the *paramātmā*, or the ultimate causal agent, is the reservoir of all form and all processes, we are merely carrying this departure to its logical conclusion. This is also consistent with the requirement that each irreducible quantum of consciousness must possess the innate capacity to comprehend and appreciate complex material forms. It stands to reason that universal consciousness should have this same capacity on a universal scale. Thus, we propose that the *paramātmā* possesses universal knowledge, and is thus able to manifest all phenomena without depending on chance or arbitrary initial conditions.

Consciousness and Knowledge

Thus far we have described the content of the consciousness of individual living entities as being limited to the perception of arrangements of matter. This content might be more or less elaborate depending on the condition of the body occupied by the conscious entity. However, it is natural to ask whether or not such a conscious entity can directly perceive either itself, other *ātmas*, or the universal consciousness, *paramātmā*.

This natural possibility opens up an entirely new avenue of practical investigation and is the most significant consequence of the view of reality which we have been outlining. Not only does it open up a new line of inquiry into the nature of life and consciousness, but it also suggests a possible method of acquiring knowledge that is different in principle from the procedure of trial and guesswork employed in science: if the absolute cause underlying all phenomena is conscious, then one might hope to obtain knowledge from this source.

The existence of higher order laws and entities incapable of mathematical description makes it completely unrealistic to hope that comprehensive knowledge of life (or matter) can be attained solely by trial and error. This can be seen very easily if we reflect on the arduous struggles which led to the formulation of the physical laws illustrated in Figure 25. How long might it take to develop an equation involving 20 times as many terms and requiring as many revolutionary changes in fundamental conceptions of nature? Yet, these very considerations also point to an alternative source of knowledge.

Let us consider how the nature of the *ātma* can be directly studied. An entity is normally studied in a scientific experiment by means of some procedure which isolates the entity in its pure and original form and eliminates extraneous influences. Such a procedure must take advantage of the particular distinguishing features of the entity and its laws of interaction.

For example, a beam of essentially pure "electricity" is generated by the apparatus known as a cathode ray tube, thus enabling many of the properties of electricity to be studied directly. This apparatus takes advantage of the basic property of electric charge and the laws of electrical interaction.

The study of the *ātma* similarly requires some procedure for isolating it in its pure state. In our normal experience the *ātma* is intimately bound up with matter by very powerful interactions, and therefore it is very difficult to discern its characteristic properties. In order to isolate the *ātma* from the influences of material interaction it is necessary to take advantage of its basic distinguishing property—consciousness—and the agency—*paramātmā*—governing its interactions. This requires the study of the relation between the individual conscious entity and the all-pervading absolute consciousness.

Thus, we cannot expect to study the *ātma* using the familiar physical laws of interaction known to modern science. The *ātma* cannot be expected to interact according to the laws of electromagnetism, and thus we could not expect to "see" one by means, say, of an electron microscope.

Rather, the isolation of the *ātma* from the influences of matter requires procedures in which one's own self becomes the object of study and the primary features and characteristics of the self are invoked. Such systematic procedures have already been extensively studied, although they tend to be relatively unknown in the cultural tradition in which our modern scientific knowledge has been developed. Their basic principles are discussed very concisely, for example, in the ancient Sanskrit text known as the *Bhagavad-gītā*.[5]

This body of knowledge may be summarized as the science of self-realization, for it is logical, consistent, and systematically related with reproducible procedures of empirical observation. We have drawn upon a small part of this science in our presentation of the concepts of the *ātma* and the absolute truth, or *paramātmā*. We have tried to show the philosophical and scientific soundness of this system of ideas as a solution to the fundamental dilemmas faced by our present scientific picture of reality.

A detailed presentation of this science is beyond the scope of this paper. We would simply like to suggest that the approach to the study of consciousness which we have outlined here will lead to new insights in our understanding of both life and matter. On the other hand, we feel that the present scientific view of life as the electromagnetic interaction of certain molecules is quite wrong and will only lead to a frustrating dead end.

Bibliography

(Chapter VII)

A. C. Bhaktivedanta Swami Prabhupada. *Bhagavad-gītā As It Is.* New York: Collier Books, 1972.

Bohm, D. *Causality and Chance in Modern Physics,* van Nostrand, 1957.

Edwards, P., ed. *The Encyclopedia of Philosophy,* Vol. 5. New York: Macmillan, 1967.

Turing, A.M. "Computing Machines and Intelligence." *Mind,* Vol. 59, 1950, pp. 433-60.

Wigner, E.P. "Physics and the Explanation of Life." *Foundations of Physics,* Vol. 1, No. 1 1970, pp. 35-45.

5 A. C. Bhaktivedanta Swami Prabhupada, *Bhagavad-gītā As It Is.*

CHAPTER 2

The Nature of Biological Form

Originally published in:
Bhaktivedanta Institute Bulletin, Volume 1, Number 1
January-February 1979, pp. 1, 4–5, 10

One of the most fundamental ideas in modern evolutionary biology is that the physical structures of living organisms can change from one kind into another through a series of small modifications, without departing from the realm of potentially useful forms. For example, the foreleg of a lizard can, according to this principle, be gradually transformed into the wing of a bird, and the lizard's scales can be gradually converted into feathers. In the course of this transformation, each successive stage is required to serve a useful function for the organism in some possible environment. Thus each intermediate form between leg and wing must be able to act as a serviceable limb under some appropriate circumstances.

The Darwinian theory of evolution is based on the hypothesis that, without exception, all the organisms in the world today came about by transformations of this kind , starting with some primitive ancestral form. If such transformations are always possible, then the problem of evolutionary theory is to determine what events in nature might cause them to actually take place. However, if there exist any significant structures in living organisms that cannot have developed in

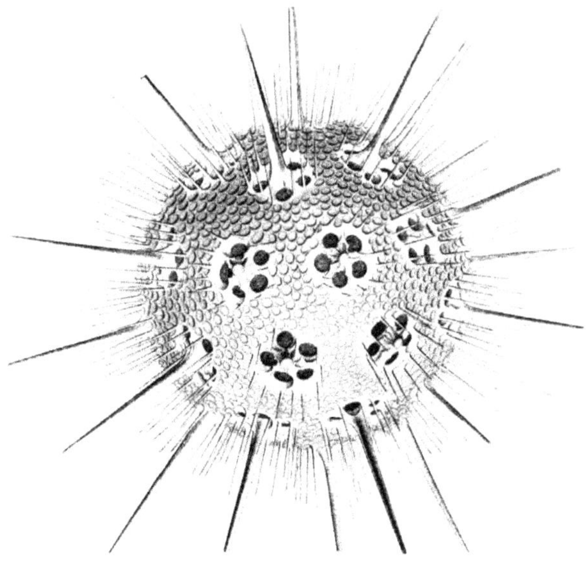

Figure 1. The radiolarian *Haeckeliana darwiniana,* a one-celled marine protozoan.

this way, then for these structures, at least, the hypothesis of evolution is ruled out, and some other explanation of their origin must be sought. Charles Darwin. the founder of the modern theory of evolution. clearly recognized this point: "If it could be demonstrated that any complex organ existed which could not possibly have been formed by numerous, successive, slight modifications. my theory would absolutely break down."[1]

Although Darwin admitted that he could not imagine the intermediate, transitional forms leading to many different organs, he assumed that they might later be revealed by a deeper understanding of the organs' structure and function, and he proceeded to base his theory on their presumed existence. However, in the nearly one hundred twenty years since the publication of his book *On the Origin of Species*, practically no significant advance has been made in the understanding of intermediate forms. While evolutionists often speak of changes in the size and shape of existing organs, they still can do very little but make vague suggestions about the origin of the organs themselves.

The geneticist Richard Goldschmidt once gave a list of seventeen organs and systems of organs for which he could not even imagine the required transitional forms. This list included hair in mammals, feathers in birds, the segmented structure of vertebrates, teeth, the external skeletons and compound eyes of insects, blood circulation, and the organs of balance.[2] These organs, and many others, present a fundamental question: How can we explain the origin of a complex system that depends on the action of many interdependent parts?

We would like to suggest here that for many organs, the reason why the required chains of useful intermediate forms are unimaginable is simply that they do not exist. Let us try to visualize this in mathematical terms. The class of all possible forms made from organic chemicals can be thought of as a multidimensional space in which each point corresponds to a particular form. We propose that in this space the potentially useful structures will appear as isolated islands surrounded by a vast ocean of disjointed forms that could not be useful in any circumstances. Within these islands some freedom of movement will exist. Corresponding to simple variations in characteristics such as size and shape. But reaching an island – corresponding to the evolution of a particular type of useful organ – will require a long and accurate jump across the ocean.

These ideas are illustrated by a mechanical example. Consider the space of all possible combinations of mechanical parts. such as shafts, levers, and gears. These mechanical parts are comparable to the molecules making up the organs in the bodies of living beings. Since both mechanical parts and molecules fit together in very limited and specific ways, a study of mechanical combinations should throw some light on the nature of organic forms.

If we visualize the space of mechanical forms. we can see that some regions in this space will correspond to wristwatches and other familiar devices. and some regions will correspond to machines that are unfamiliar, but that might function usefully in some situation. However, the space will mostly consist of combinations of parts that are useful as paperweights, at best.

Since a machine can operate smoothly only if many variables are simultaneously adjusted within precise limits, the useful machines will occupy isolated islands, surrounded by an ocean of machines that are jammed or broken. If we started from a point on the shore representing a very rudimentary machine, or no machine at all, then we would have to leap over this ocean in order to reach, say, a functional wristwatch. As we made this leap, we could not obtain any guidance by testing the relative usefulness of the forms beneath, for all of them would be equally useless.

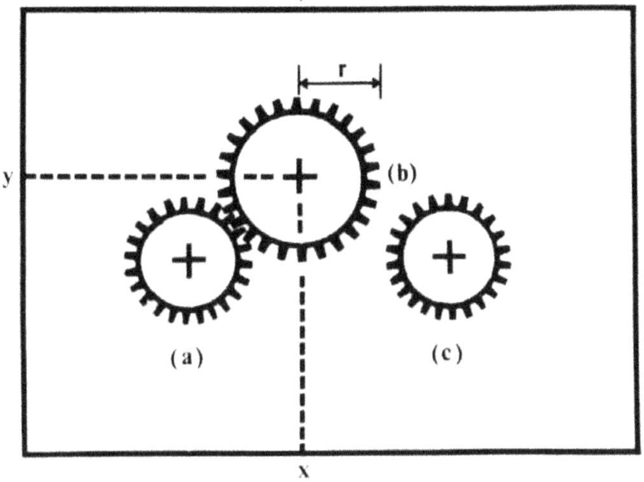

Figure 2. A simple mechanical system that shows how several variables must interact in a precise fashion to produce a useful biological form.

A very simple mechanical arrangement can be used to show how natural constraints on the combination of parts limit the class of functional machines. Figure 2. depicts an arrangement of three gear wheels – (a), (b), and (c) – within a box. The situation of wheel (b) can be partially described by specifying its radius, r, and the x and y coordinates of its center. If we allow r, x, and y to vary, we obtain a three-dimensional space of possible configurations for this mechanical system. This is depicted in Figure 3.

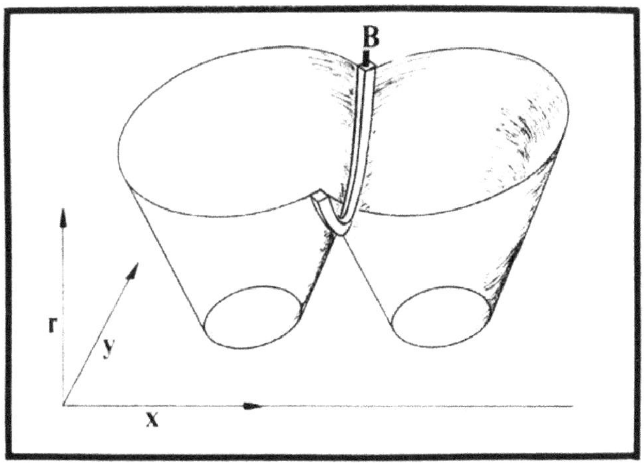

Figure 3. The space of possible configurations of [he wheel (b) in Figure 2. The thin strip B represents the configurations in which wheel (b) meshes with both (a) and (c).

In this space there are two surfaces shaped like inverted cones. The cone on the left represents the constraints on r, x, and y necessary for wheel (a) to engage wheel (b), and the cone on the right represents the similar constraints for wheels (b) and (c). (These cones should possess a certain thickness, representing the limits within which the gears will mesh.) The region B, where the two cones intersect, represents the island of configurations where (b) meshes with both (a) and (c). Power can be transmitted from (a) to (c) only in these configurations.

In general, functional configurations of parts will be represented by the intersection in a multidimensional space of many surfaces, corresponding to various constraints on the combination of the parts. This should also hold true when the parts are the molecular components forming the bodies of living organisms. As the number of components is increased, both the dimensionality of the space and the number or constraining surfaces will tend to increase. This can be expected to lead to increasing isolation for the islands representing functional organs.

Since the bodies of living organisms are very complicated, they are not as easy to visualize as the machines in our illustration. However, there are examples of organs that are simple enough to be comparable with man-made mechanisms. One such example is found in the one-celled bacterium *Escherichia coli* (Figure 4).

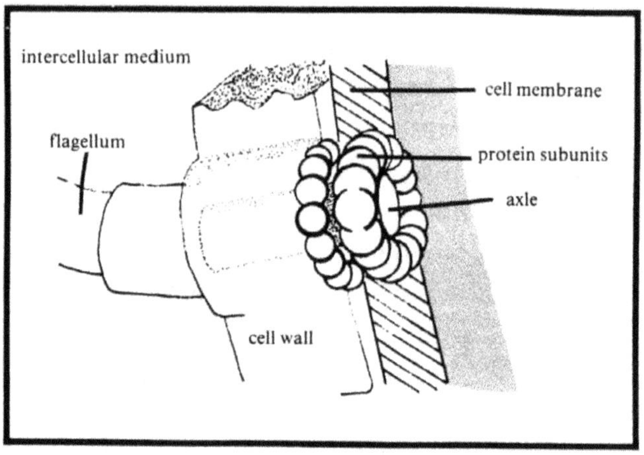

Figure 4. A schematic depiction of an Escherichia coli's flagellar motor.

Each *Escherichia coli* cell possesses several long, curved fibers (called flagella) that enable it to swim.[3] Each flagellum is connected alone end to a kind of motor built into the bacterial cell wall, and when these motors rotate in a certain direction the flagella rotate in unison and act as propellers to drive the bacterium forward through the water. When the motors rotate in the opposite direction, the flagella separate and change the orientation of the bacterium by pulling in various ways. By systematically alternating between these two modes of operation, the bacterium is able to swim from undesirable to desirable regions of its environment.

The motors are presently thought to be driven by a flux of protons flowing into the cell. Each motor is thought to consist of a ring of sixteen protein molecules attached to an axle and a stationary ring of sixteen proteins built into the cell wall.[4] Protons are steadily pumped out of the cell by its normal metabolic processes. As some of these protons flow back into the cell through the pairs of rings, they impart a rotary motion to the movable ring. Since the motor can operate in forward or reverse, there must be some mechanism that adjusts the molecules in the rings so as to reverse the direction of rotation.

Although the exact details of the *Escherichia coli's* molecular motors have not been worked out, we can see that they depend on the precise and simultaneous adjustment of many variables. In the space of possible molecular structures, the functional motors will represent a tiny, isolated island. To have a continuum of useful forms spanning the gap between "no motor" and "motor," we would have to postulate useful organs that do not function as motors but are very similar to motors in structure. For the selective processes of evolutionary theory to eventually choose a working motor, these non-motors would have to be progressively more useful to the bacterium the more motor-like they became. Apart from this very unlikely possibility, evolutionists can suggest no guiding process that can cross the gap.

In the case of very simple organs, such as the bacterial motor, it should be possible to carry out a completely rigorous study of the possibilities of form. Such a study would definitely resolve the question of whether the intermediate forms required by the theory or evolution do or do not exist. Of course, for the highly complicated organs of higher plants and animals, this kind of study may not be practical, but there are still many cases where the combinatorial logic of an organ strongly suggests the impossibility of useful intermediate forms.

One interesting example of this impossibility is found in the statocyst of a certain species of shrimp.[5] The statocyst is a hollow, fluid-filled sphere built into the shrimp's shell. It is lined with cells bearing pressure-sensitive hairs and containing a small weight. The weight tends to sink and press against the downward portion of the sphere, thus enabling the shrimp to tell up from down. Curiously, the weight is a small grain of sand that the shrimp picks up with its claws and inserts into the statocyst through a small hole in its shell. The shrimp has to do this every time it molts its shell.

Now, the question is this: By what intermediate stages did the arrangement of the shrimp's statocyst come about? Both the statocyst and the behavioral pattern involved in picking up the grain of sand are quite complex, and neither is of any use without the other. Even if a statocyst evolved with a built-in weight and then lost this feature by a mutation, the appearance of the insertion behavior would require a leap involving the coordination of many variables.

At this point, let us try to find an alternative explanation of how such a leap might come about. One natural process in which such leaps are commonly seen is the process of human invention. The products of human creativity, from watches to poetic compositions, are generated with the aid of inspiration, which enables one to proceed directly to the solution of a problem without groping laboriously through many false attempts. In fact, it is often the case that after experiencing great frustration in a totally futile trial-and-error search, an inventor will see the complete solution to his problem in a sudden flash of insight. One example of this is the experience of the mathematician Carl Gauss in solving a problem that had thwarted his efforts for years: "I succeeded," he wrote, "not on account of my painful efforts, but by the grace of God. Like a sudden flash of lightning, the riddle happened to be solved. I myself cannot say what ... connected what I previously knew with what made my success possible."[6] It is significant that the solution did not exhibit even a hint of a connection with Gauss's previous attempts. Here again we find a structure – this time a structure of abstract thought – that is not linked by any discernible chain of intermediate forms to other, existing structures.

If it is in the nature of biological form and the forms of human invention to exist as isolated islands in the sea of possible forms, then some causal agency must exist that can select such forms directly. The experience of inventors indicates that this agency lies outside the realm of human consciousness or control, and that it is capable of acting very quickly.

In the *Bhagavad-gītā* a unified description is given of an agency that accounts for the origin of both biological form and human creativity. There it is explained that the ultimate cause underlying the world of our perceptions is not a blind, impersonal process, but a primordial, absolute personality – a personality possessing eternal form, qualities, and activities. Thus, in the *Bhagavad-gītā* Śrī Kṛṣṇa declares, "I am the father of all living entities" (Bg. 14.4), and also "I am seated in everyone's heart, and from Me come remembrance, knowledge, and forgetfulness" (Bg. 15.15).

Of course, even though intermediate biological forms commonly do not exist (implying some kind of absolute information or guidance that transcends the categories of ordinary science), this is not sufficient in itself to bring us to the conclusion that the transcendental source must be the Supreme Person. However, this hypothesis opens up very interesting opportunities for further scientific investigation. If the information for the manifestations of form and order in this world is existing in a transcendental state, then this information might be directly accessible in some way. And if the transcendental source is indeed the Supreme Person, as described in the *Bhagavad-gītā*, then it is reasonable to expect that a personal avenue of approach is possible.

In fact, such an avenue does exist. It consists of an elaborately developed scientific method for establishing a personal relationship with the Supreme

Person. This method, called *bhakti-yoga*, is similar to modern science in that it depends on clearly specified procedures leading to reproducible results. It is fully empirical, for it is based on direct personal experience that is attainable by anyone who carries out the procedures correctly.

On the other hand, *bhakti-yoga* differs from modern science in its method of acquiring basic information. In modern science the hypotheses to be tested, as well as the methods for testing them, are obtained in a haphazard way from the poorly understood sources of "inspiration," or "creative imagination." In the science of *bhakti-yoga*, experimental procedures and philosophical principles are both explicitly obtained from the Supreme Person. In other words, although the source of knowledge in both modern science and *bhakti-yoga* is the Supreme Person, in *bhakti-yoga* this is fully recognized. and thus there is direct access to the transcendental knowledge available from this source. A good example of this direct access is the *Bhagavad-gītā* itself, which, far from being a product of gradual cultural evolution, was directly spoken by Śrī Kṛṣṇa some five thousand years ago on the battlefield of Kurukṣetra.

It would be worthwhile for scientists to consider this direct method of attaining knowledge. Even though history has shown that revealed knowledge may become corrupted, the basic principle is still valid, and fruitful scientific investigation in this area should be possible. The value of seeking such a rigorous approach is especially apparent if, as we have seen, there is reason to suppose that organized form in both the biological and cultural spheres must originate from a transcendental source.

References

1. Charles Darwin, *The Origin of Species* (New York: D. Appleton and Co., 1898), p. 229.
2. Richard Goldschmidt, *The Material Basis of Evolution* (New Haven: Yale Univ. Press, 1940), pp. 6–7.
3. Howard C. Berg, " How Bacteria Swim," *Scientific American* (Vol. 233, No.2, 1975), pp. 36--44.
4. Peter C. Hinkel and Richard E. McCarty, "How Cells Make ATP," *Scientific American* (Vol. 238, No. 3, 1978), p. 116.
5. W. von Buddenbrock, *The Senses* (Ann Arbor: Univ of Michigan Press, 1958), pp. 138–141.
6. Jacques Hadamard. *The Psychology of Invention in the Mathematical Field* (Princeton: Princeton University Press, 1949), p. 15 .

CHAPTER 3

Numerical Analysis and Theoretical Modeling of Causal Effects of Conscious Intention

Originally published in:
Subtle Energies and Energy Medications
Volume 2, Number 1 (1991), pp. 47–70

Abstract

In this paper we discuss some phenomena involving interactions between machines and states of conscious intention that have been reported by Robert Jahn and his colleagues at Princeton University. We specifically deal with the experiments carried out by these investigators with an apparatus called the random mechanical cascade (or RMC). We introduce a class of theories, called selection theories, which might be invoked to explain the phenomena they have observed. These include some parapsychological theories that have been proposed for such phenomena in the past, and they also include the theory that the phenomena are spurious by-products of conscious or unconscious editing of the experimental data.

We have found that the data for the RMC experiments have some statistically significant features which have not been noted before. and which tend to rule out selection theories as possible explanations of the observed phenomena. Thus our findings support the conclusion that these phenomena represent a genuine anomaly, and they narrow down the range of possible theories that might account for this anomaly.

Introduction

A group at Princeton University, including R. D. Nelson, B. J. Dunne, and R. G. Jahn,[1] has for several years carried out a research program investigating correlations between human intentions and the behavior of various machines. In one series of experiments the machine was a version of a device commonly used to demonstrate the law of large numbers. The Princeton device, called a "Random Mechanical Cascade" or RMC, was described as follows in Nelson, *et al.*[2] The machine consists of a quincunx array of 330 ¾" nylon pins with a horizontal spacing of 3.25". A total of 9000 ¾" polystyrene spheres are allowed to cascade through the array from an inlet at the top, and these accumulate in 19 equally spaced collecting bins at the bottom. About 12 minutes are required for all of the balls to reach the bins.

As one would expect, the 9000 balls tend to fill the bins according a *Gaussian* distribution, but this distribution is somewhat irregular, and its mean and other statistical features tend to vary randomly. In the Princeton experiments an observer, called an "operator," would try to influence the mean of the distribution to shift to the left, remain at the baseline (the statistically expected position), or shift to the right. The operator did this by meditation within the mind, rather than by trying to physically interfere with the machine. Usually the operator sat in front of the machine at a distance of about eight feet and watched the cascading balls, but in some experiments the operator was at a remote location.

The operation of the machine was divided into sets of three "tripolar" runs, one for each of the three intentions of left, baseline, or right. (In some cases the operator was free to choose the order of the intentions in each set, and in other cases this was dictated by the experimental protocol.) An operator would perform a number of series, each consisting of 10 or 20 tripolar sets. The distribution of the balls in the bins was counted electronically for each run and recorded automatically in a computer file.

For a given run, let $b(k)$ be the number of balls in bin k for $k = 1, \ldots, 19$. The 19 numbers, $b(1), \ldots, b(19)$, are referred to as the bin distribution for the run, and they roughly approximate a *Gaussian* distribution. The mean of this distribution is called the bin distribution mean. We will also speak of the mean and standard deviations of the variable called "bin distribution mean" over a series of runs, each of which produces a bin distribution and a bin distribution mean. These two uses of the word "mean" can be distinguished by context.

The performance of the operators was evaluated by examining the behavior of the quantities *LT-BL*, *RT-BL*, and *RT-LT*, where *LT*, *BL*, and *RT* are the cumulative bin distribution means for the intentions, left, baseline, and right, over a long series of runs. Differences between *LT*, *BL*, and *RT* were used to offset the effect of long term trends in machine behavior caused by wear and other systematic factors. We note that since there are19 bins, the bin distribution mean for a perfectly symmetrical machine should come out to 10. Since the actual machine is slightly asymmetrical, this average tends to be slightly higher than 10.

One would naturally expect that there would be no relationship between human states of consciousness and the statistical behavior of the machine. However, the experiments indicated that in the long run, this behavior tended to conform with the intentions of the operators. In a total of 1131 tripolar sets of runs generated by 25 operators, it was found that the cumulative *RT-LT* came out to about .0057. This corresponds to an average mean for rightward intentions of 10.0229 and an average mean for leftward intentions of 10.0172.

Although the differences between these average bin means are quite small, they turn out to be statistically significant The standard deviation for *RT-LT* was .0493, and the *t*-score was 3.891. This outcome has a probability of about 5×10^{-5}, if we assume that the individual runs were statistically independent and not influenced by operator intention.

In these experiments the bulk of the runs were generated by two operators (10 and 55), and operator 10 achieved by far the most significant results. However, a statistically significant effect remains, even if the data generated by operators 10 and 55 are excluded from the analysis.

In this paper we will analyze the results of the random mechanical cascade experiments, and show that the data generated by these experiments reveals some anomalous effects in addition to those originally reported. To do this, it was necessary for us to gain access to the original RMC data, and this was kindly provided by Roger Nelson of the engineering anomalies research group at Princeton.[1] In Section 1 we briefly discuss the the possible effects of small forces generated by the observer on the RMC. In Section 2 we introduce the class of selection theories and give three examples of such theories. In Section 3 we discuss the RMC data, and we show how it reveals hitherto unknown effects that should not occur, according to selection theories. Finally, in Section 4 we summarize our conclusions.

1. The Effect of Small Forces on the RMC

What happens when a conscious person desires to perform some action, such as picking up an object? Standard explanations maintain that the desire to act can be identified with particular electro-chemical phenomena in neurons within the brain. These give rise to movements of bodily parts and possible changes in the body's electrical field and other physical characteristics. Could these changes in turn influence the behavior of the RMC in a normal physical fashion?

Nelson *et. al.*[3] discuss possible mechanical, electromagnetic, and gravitational interactions between the observer and the RMC, and reject these "ordinary" forces as being too weak by several orders of magnitude to produce the observed effects. Although this is basically correct, there are some fine points regarding the effect of ordinary forces on the RMC that should be taken into account.

In recent years a phenomenon known as deterministic chaos has been extensively investigated.[4] The essence of this phenomenon is that in physical systems with nonlinear dynamics, very small changes imposed on the state of the system can grow exponentially into very large changes in a short period of time. The consequences of this exponential growth have been dramatically illustrated by the "butterfly effect," described by E. N. Lorenz in a meteorological context: "even if the atmosphere could be described by a deterministic model in which all parameters were known, the fluttering of a butterfly's wings could alter the initial conditions, and thus (in the chaotic regime) alter the long term prediction."[5]

The dynamics of the RMC involve many nonlinear interactions, such as the nonlinear dependence of angle of bounce on angle of incidence in collisions between balls and pins. Analysis of simple mathematical models of the RMC shows that these interactions result in deterministic chaos, and we suggest that they also result in chaotic phenomena in the real RMC. If this is true, it means that extremely small forces generated within the environment of the RMC could have a measurable effect on the bin distribution mean. Thus, in principle, the bin distribution mean could be influenced in a systematic way by small, ordinary, physical forces generated by the body of the observer.

However, if small forces can result in large effects, small variations in these forces can likewise result in large variations in the effects. According to the theory of deterministic chaos, this means that if the observed anomalous behavior of the bin distribution mean is produced by small, ordinary forces generated within the body of the observer, then these forces must be produced with great precision in accordance with the observer's intentions. Thus as Jahn and his colleagues note,[1] the observed anomalous effects cannot be produced by normal forces applied with a normal degree of precision (of the kind we would expect from a human nervous system). But they might be produced by normal forces applied with paranormal, or unexpectedly high, precision. Here we will not try to quantify "paranormal precision", but we suggest that it might be worthwhile trying to do this in future studies of PK phenomena.

2. Selection Theories

Of course, there are theories which attempt to account for the anomalous behavior of the RMC without bringing in physical forces at all. For example, there is a class of theories, called selection theories, that can be described in general terms as follows: Let us suppose that natural processes obeying known physical laws produce an ensemble of possible outcomes for runs of the RMC. This ensemble can be defined by a probability distribution, $P(x)$, where x is the information representing an RMC run. Let y represent the observer's state of intention, and let $f(x, y) > 0$ measure the degree to which a run x satisfies the intention y. (Here we write $y = RT$, LT, or BL, where these symbols stand for states of intention.) We suppose that the greater the agreement between the run and the intention, the greater is $f(x, y)$. Define the following probability distribution for a pair, (x, y). consisting of a run plus an intention:

$$P'(x,y) = f(x,y)\, P(x) / K \qquad (1)$$

where K is a normalization constant. In a selection theory, the actual, measured probability distribution for (x, y) is given by $P'(x, y)$ for some suitable function, $f(x, y)$. We will give three examples of selection theories, and show in each case how one arrives at $P'(x, y)$.

Example (1), precognition. The first example assumes that the RMC always behaves according to known physical laws, but that the observer has a paranormal ability to foresee the future. In this example we suppose that the observer has the opportunity to make his own choice of intentions prior to each run (the "volitional" mode). We suppose that he makes his choice in accordance with what he foresees, but that his foresight is imperfect This can be expressed by means of a Markov chain transition matrix, $M(x, y)$, which gives the probability that the observer will choose intention y, given that the future run will be x. In this model, the probability that (x, y) will come up is

$$P'(x,y) = P(x)\,M(x,y) \qquad (2)$$

which has the same form as Equation (1). The constant $K = 1$ since the sum of $M(x,y)$ over $y = RT, BL, LT$ is assumed to be equal to 1.

In the case where the intentions are chosen in advance by the experimenters (the "instructed" mode), a similar model can be formulated on the basis of precognition by the experimenters.

Example (2), wave function collapse. We have studied the behavior of solutions of the Schrödinger equation for an idealized model of balls bouncing from pins. The results suggest that during a run, the quantum mechanical wave function for the RMC should spread out to encompass a range of outcomes for that run. Thus, the normal physical behavior of the RMC can be expressed in terms of the collapse of the quantum mechanical wave function. We can let Ψ represent a wave function of the RMC plus observer that encompasses many possible run outcomes, and let Φ_i represent wave functions of the RMC plus observer for specific outcomes. (Just to simplify the discussion, let us suppose that the Φ_i's form an orthonormal basis for wave functions.)

In quantum mechanics, the probability of getting state Φ_i after collapse of the wave function, Ψ, is $|<\Psi, \Phi_i>|^2$. Wigner[6] has argued that the collapse of the wave function is connected with consciousness, and he proposes that a proper formulation of this connection will require modifications of the laws of quantum mechanics. So, if we suppose that consciousness prefers Φ_i's showing harmony between intention and RMC run, we can venture to express this by replacing $|<\Psi, \Phi_i>|^2$ by $f(\Phi_i)|<\Psi, \Phi_i>|^2$, where f is a positive function that favors such harmony. This modified form of quantum mechanics is also a selection theory.

Example (3), data selection. In this theory we suppose that the observed correlation between intention and RMC behavior is spurious. The probability of run outcome x is determined by a distribution $P(x)$, which is generated in accordance with the laws of physics. But in the natural course of events, some runs may seem defective for one reason or another, and the experimenter may wish to delete them from the database. We suppose that, owing to the desire for a successful experiment, the experimenter is more likely to throw out runs where x disagrees with the intention, y, than runs where x agrees. (He may do this consciously or subconsciously.) This can be expressed by means of a data selection function, $0 < f(x,y) < 1$, which represents the chances of retaining a particular run in the database, and which is somewhat smaller for cases where x and y disagree than it is for cases where they agree. The probability for (x, y) in the edited database is then given by Equation (1).

Let us suppose that the run information, x, can be expressed as $x = (u, v, w)$, where u and v are real variables describing different features of the run, and w contains whatever additional information is needed to specify x. We can suppose that v is the variable to which intentions apply (namely, the bin mean in the RMC experiments). Since $f\ (x, y)$ expresses the degree of agreement between v and the intention, y, we can write it as $f_1\ (v, y)$. The variable u is presumed to be one to which the intentions of the operator do not apply; for example, this would be the case if the operator had no knowledge of u.

If we substitute $x = (u, v, w)$ into Equation (1), and then eliminate the variable w by summing over it, we obtain an equation of the form

$$P'\ (u, v, y) = f_1\ (v, y)\ P\ (u, v)\ /\ K \qquad (3)$$

The dependence of v on y in this model can be seen by examining the expected value of v, given y. This expectation value is

$$E\ (v \mid y) = \int\int v\ P'\ (u, v, y)\ dudv\ /\ \int\int P'\ (u, v, y)\ dudv \qquad (4)$$

and we can similarly define $E\ (v \mid y)$, the expected value of u, given y.

The model is designed so that $E\ (v \mid y)$ varies as the intention, y, varies. But what does the model say about how the variable, u, varies as y varies? We can readily calculate this if we assume that the distribution, $P\ (u, v)$, for u and v is a bivariate normal distribution with means, m_u , m_y , standard deviations, s_u , s_v , and correlation coefficient, r. Given this assumption, we find that,

$$[E\ (v \mid y) - m_u]\ /\ s_u = r\ [E\ (v \mid y) - m_v]\ /\ s_v \qquad (5)$$

The distribution, $P\ (u, v)$, depends on the physics of the RMC. This equation tells us that if u and v are not strongly correlated in $P\ (u, v)$, then $E\ (u \mid y)$ will vary only weakly as the intention, y, is varied. This makes sense because u is not directly linked to the intention, y. It is linked with y only indirectly through $f_1\ (v, y)$ and $P\ (u, v)$.

We will show in the next section that we can find an RMC variable, u, for which $P\ (u, v)$ is approximated by a bivariate normal distribution with a small r, but for which $E\ (u \mid y)$ does vary strongly with y, in violation of Equation (5). Indeed, even though r is positive, $E\ (u \mid y)$ and $E\ (v \mid y)$ vary in opposite directions as y varies. This implies that the RMC data reported by Nelson *et al.*,[1] cannot be realistically modeled by a selection theory, and in particular, it cannot be modeled by theories (1) through (3), above.

3. Analysis of the Data

The information describing an RMC run is given by 19 bin numbers, b (1), . . . ,b (19), indicating the number of balls that fall in each of the 19 bins. As we mentioned in the introduction, this sequence of 19 numbers is called the bin distribution for the run. The bin mean is simply the expected value of *i* for the distribution, *b* (*i*), i = 1, . . . ,19. We wanted to find additional variables that (1) are functions of the 19 bin numbers, (2) are physically meaningful, and (3) are nearly statistically independent of the bin mean and of one another. Our objective in defining such variables was to find candidates for the *u* of Equation (5).

The distribution of balls in the bins can be approximated by a *Gaussian*. In Nelson *et. al.*[7] it is observed that the bin distribution tends to deviate from a *Gaussian* in the center due to the tendency of some balls to fall straight down for some distance without striking pins, and at the sides due to bouncing of the balls from the sides of the machine. The deviation in the center takes the form of a second, narrower *Gaussian* superimposed on the main bin distribution *Gaussia*n. The bouncing of the balls from the sides contributes an additional U-shaped curve that is superimposed on these two *Gaussians*. Thus, the bin distribution can be broken down into the sum of these three curves, each of which can be explained in terms of the physics of the RMC.

Let *Dist*(1), . . . , *Dist*(19) stand for the bin distribution for a given run. We tried to break down *Dist* into the three components just mentioned. The first of these is *Gauss*1, an approximation to *Dist* by a *Gaussian*. This was obtained by fitting a quadratic to log [*Dist*(3), . . . , log [*Dist*(7)], log [*Dist*(13)], . . . , log [*Dist*(17)] by least squares, and expressing *Gauss*1 as the exponential of that quadratic. In the least squares fit, we avoided the bins in the middle and on the ends since for these bins, *Dis*t tends to deviate from a *Gaussian*.

The next step is to scale *Gauss*1 so that its total, *Gauss*1 (1)+ . . . +*Gauss*1 (19), is maximal, given that

$$Dist\,2\,(\,i\,) = Dist\,(\,i\,) - Gauss1\,(\,i\,) \geq 0 \qquad (6)$$

for *i* = 1, . . . ,19. When we do this we find that *Dist* 2 has the shape of a *Gaussian*, but is narrower than *Gauss*1. We therefore define the second component as *Gauss*2, a *Gaussian* approximation to *Dist* 2. This *Gaussian* is defined to have the same mean and standard deviation as *Dist* 2 on bins 3–17 (avoiding the non-*Gaussian* behavior on the sides), and it is scaled so that its root-mean-square difference from *Dist* 2 is minimal.

The total of *Gauss*1 cannot exceed 9000, the total number of balls per run, and it generally is about 7700. The total for *Gauss*2 is generally somewhat over 1000. The third component. called *Rem* (for remnant), is defined by

$$Rem\ (\ i\ ) = Dist\ 2\ (\ i\ ) - Gauss2\ (\ i\ ) \qquad (7)$$

for $i = 1, \ldots, 19$. We note that *Rem* (*i*) may take on negative values.

Figure 1 illustrates the breakdown of *Dist* into *Gauss*1, *Gauss*2, and *Rem* in a particular case. It is interesting that *Rem* has a remarkably consistent form.

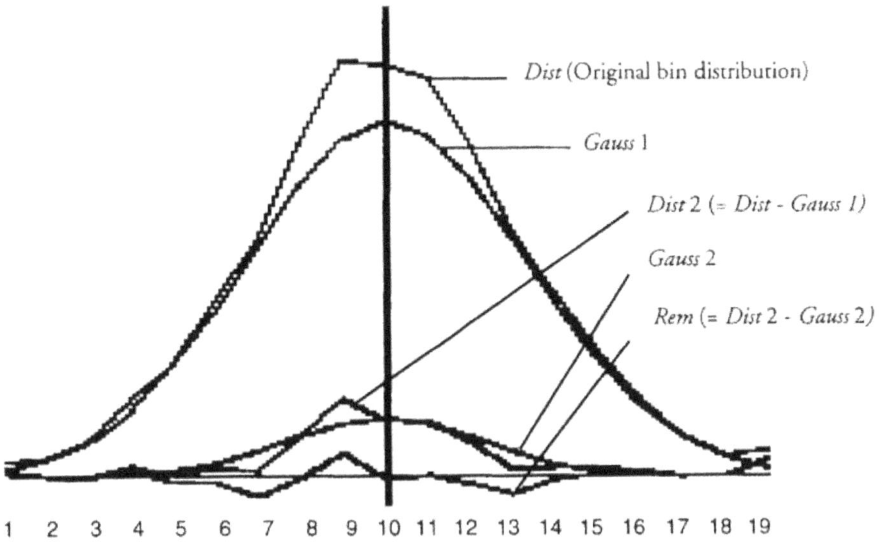

Figure 1. Breakdown of bin distribution into component curves.

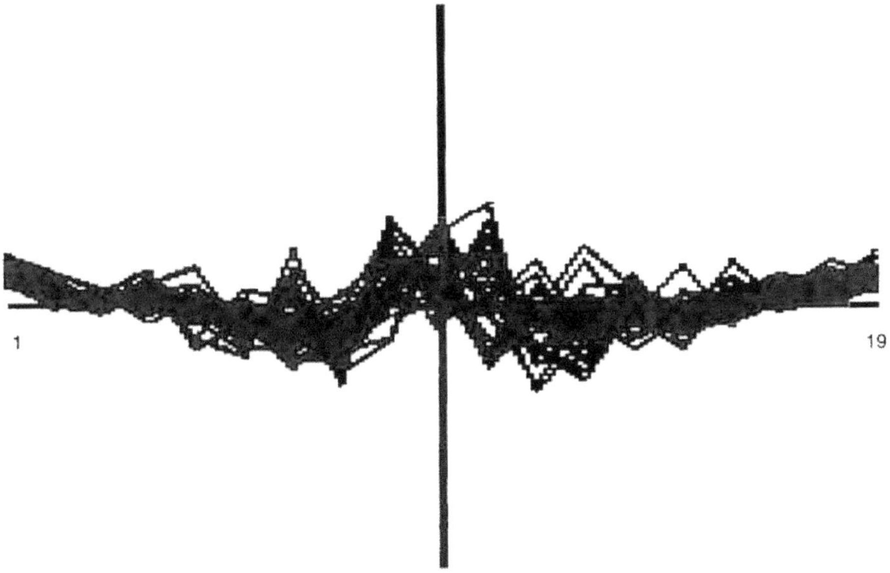

Figure 2. Superposition of 30 examples of Rem (= Dist - Gauss1 - Gauss2).

For each run, we can compute a number of standard parameters for the distributions, *Dist*, *Gauss*1, and *Gauss*2, including their totals, means, and standard deviations. We can also compute such quantities as *Rem* (1) + *Rem* (2) and *Rem* (18) + *Rem* (19), which indicate the behavior of the bouncing balls near the sides of the RMC.

Table I

A Set of 14 bin distribution variables,
based on *Dist*, *Gauss*1, *Gauss*2, and *Rem*

Vble.	Mean	S.D.	Description
V1	10.0082	.2449	Mean of *Dist-Gauss*1
V2	10.0341	.0774	Mean of *Dist* restricted to bins 3-7,13-17
V3	115.9227	11.8348	*Dist* (18) + *Dist* (19)
V4	10.0178	.0382	Mean of *Dist*
V5	10.0194	.0528	Mean of *Gauss* 1
V6	10.0480	.2120	Mean of *Gauss* 2
V7	3.2750	.0383	S.D. of *Dist*
V8	38.3002	12.2195	*Rem* (18) + *Rem* (19)
V9	3.3052	.0684	S.D. of *Gauss* 1
V10	2.6706	.2378	S.D. of *Gauss* 2
V11	7774.6040	443.9916	Total of *Gauss* 1
V12	1157.4250	447.0527	Total of *Gauss* 2
V13	67.9623	27.9438	Total of *Rem*
V14	42.3580	12.3441	*Rem* (1) + *Rem* (2)

Figure 2 shows the superposition of *Rem* curves obtained from 30 different bin distributions. The elevation at the sides is presumably due to reflection of balls from the side walls, and the other characteristics must be due to asymmetries in the structure of the machine.

For each run, we can compute a number of standard parameters for the distributions, *Dist*, *Gauss*1, and *Gauss*2, including their totals, means, and standard deviations. We can also compute such quantities as *Rem* (1) + *Rem* (2) and *Rem* (18) + *Rem* (19), which indicate the behavior of the bouncing balls near the sides of the RMC.

Table I gives a list of 14 of these quantities, called *v* 1, ... ,*v* 14, along with their means and standard deviations over our entire set of RMC data, which consists of 4530 (= 3 × 1510) runs. Table II lists the correlation coefficients for pairs of the variables *v* 1, ... ,*v* 14. These are also computed using the entire data set.

Table II

Statistical correlations over the total data set for variables *v* i, ..., *v* 14
The total data set contains 4530 bin distributions.

	2	3	4	5	6	7	8	9	10	11	12	13	14
1	-.11	.29	.18	-.46	.85	-.04	.38	-.02	-.01	-.04	.04	.01	-.33
2		.05	.89	.85	.00	.08	-.13	-.01	-.09	-.01	.02	-.04	.18
3			.28	.03	.01	.53	.82	.22	-.11	.04	-.06	.27	.05
4				.74	.12	-.10	.12	-.03	-.08	-.04	.04	-.04	-.08
5					-.42	-.07	-.17	-.01	-.07	.00	.00	-.05	.17
6						.00	.08	-.04	.00	.03	-.03	.02	.00
7							.10	.59	.12	.04	-.05	.16	.17
8								-.28	.05	.09	-.12	.52	-.26
9									-.41	-.45	.48	-.55	-.26
10										.03	-.08	.68	.03
11											-1.00	.09	.18
12												-.15	-.21
13													.49

We wanted to study a set of bin distribution variables that have as little statistical correlation as possible, but are also physically meaningful. One way of doing this is to eliminate strongly correlated variables from the set *v* 1, ... , *v* 14 until a set of nearly uncorrelated variables remains. Table III shows the result of doing this.

In the set of variables of Table III, *v* 4 is the bin mean, which the RMC operators tried to directly influence. The other 6 variables were presumably not the object of operator or experimenter intentions, and they would be hard for the operator to recognize by observing the machine. Our question is: Do these other variables display behavior significantly correlated with operator intention? A selection theory would imply they should not do this, but it turns out that some of them do.

Table III

A subset of the variables v I, ... , v 14 which are nearly statistically independent of one another.

		6	7	8	10	11	14
4	Mean of *Dist*	.12	-.10	.12	-.08	-.04	-.08
6	Mean of *Gauss* 2		.00	.08	.00	.03	.00
7	S.D. of *Dist*			.10	.12	.04	.17
8	*Rem*(18) + *Rem*(19)				.05	.09	.14
10	S.D. of *Gauss* 2					.03	.03
11	Total of *Gauss* 1						.18
14	*Rem* (1) + *Rem* (2)						

In Table IV we list results of an analysis similar to that reported in Tables I.A–C in Nelson *et. al.*[1] Their table was computed for a 1131 run subset of the total data set, generated by 25 out of the 35 operators. These comprise all the runs in the total data set in which a single operator observed the RMC machine in operation, in the same room. (Other runs involved multiple operators, or an operator trying to influence the machine at a remote location or at a time before or after the time of the run.) In our analysis we also made use of this 1131 run subset.

For each tripolar set of runs, the differences $v\,4\,(i, LT) - v\,4\,(i, BL)$, $v\,4\,(i, RT) - v\,4\,(i, BL)$, and $v\,4\,(i, RT) - v\,4\,(i, LT)$ were computed, where $v\,4\,(i, LT)$, $v\,4\,(i, BL)$, and $v\,4\,(i, RT)$ are the bin means obtained under leftward, baseline, and rightward intentions for the ith tripolar set.

(Here $i = 1, \ldots, 1131$). The means and standard deviations for these three differences were computed for the 1131 cases, and were used to compute corresponding t-scores and probabilities. The probabilities indicate how likely it is that the means of the differences would be displaced by chance from 0 to their observed values. The main point made by Jahn and his colleagues is that these probabilities turn out to be unexpectedly low.

Table IV

Comparison between the behavior of the bin mean, v 4, and variables *v* 5, *v* 6, *v* 7, *v* 8, *v* 10, *v* 11, *v* 14, *v* 15, and *v* 16 for 1131 runs generated by 25 operators.

Vble	Intent	Mean	S.D.	t-score	Prob.	Correlation
4	*LT-BL*	-.00564	.05005	-3.7870	.76E-04	
4	*RT-BL*	.00007	.05008	.0472	.48E+00	
4	*RT-LT*	.00571	.04931	3.8913	.50E-04	
5	*LT-BL*	-.00789	.07090	-3.7420	.91E-04	.70955
5	*RT-BL*	.00001	.06938	.0065	.50E+00	.72223
5	*RT-LT*	.00790	.06927	3.8364	.62E-04	.71710
6	*LT-BL*	.00790	.29531	1.2799	.10E+00	.09233
6	*RT-BL*	.00373	.29996	.4178	.34E+00	.12404
6	*RT-LT*	-.00751	.29278	-.8629	.19E+00	.12407
7	*LT-BL*	.00321	.03500	3.0875	1.00E-03	-.04128
7	*RT-BL*	.00085	.03578	.8003	2.1E-01	-3315
7	*RT-LT*	-.00236	.03630	-2.1881	1.4E-02	.01918
8	*LT-BL*	.70664	17.63451	1.3476	8.9E-02	16174
8	*RT-BL*	-.59116	17.25134	-1.1524	1.2E-01	.08853
8	*RT-LT*	-1.29780	16.99774	-2.5677	5.1E-03	.12033
10	*LT-BL*	-.00397	30761	-.4343	3.3E-01	.00937
10	*RT-BL*	-.01122	.31699	-1.1900	1.2E-01	.02250
10	*RT-LT*	-.00724	31543	-7724	2.2E-01	-.00912
11	*LT-BL*	-18.69107	630.79480	-9965	1.6E-01	.01230
11	*RT-BL*	-4.65074	632.71381	-0.2472	4.00E-01	.00503
11	*RT-LT*	14.03979	627.16571	7529	2.3E-01	.03185
14	*LT-BL*	1.02498	16.41335	2.1001	1.8E-02	-10295
14	*RT-BL*	-10467	16.80345	-.2095	4.2E-01	-.10887
14	*RT-LT*	-1.12965	17.39774	-2.1836	1.4E-02	-7294
15	*LT-BL*	.33160	3.08571	3.6140	1.5E-04	.05171
15	*RT-BL*	.03976	3.08411	4336	3.3E+01	.02832
15	*RT-LT*	-.29184	3.08992	-3.1764	7.5E-04	.07167
16	*LT-BL*	.27337	2.82734	3.2517	5.7E-04	.06682
16	*RT-BL*	-.01755	2.87298	-.2055	4.2E-01	.04326
16	*RT-LT*	-.29093	2.88306	-3.3936	3.4E-04	.08805

We performed the same computations for each of the remaining 6 variables of Table III, plus the additional variable, *v* 5. For example, for variable *v* 14 the computations were done for the differences, *v* 14 (*i*, *LT*) - *v* 14 (*i*, *BL*), *v* 14 (*i*, *RT*) - *v* 14 (*i*, *BL*), and *v* 14 (*i*, *RT*) - v 14 (*i*, *LT*). Table IV enables us to compare the probabilities computed for each of these 7 variables (namely, *v* 5, *v* 6, *v* 7, *v* 8, *v* 10, *v* 11, and *v* 14) with the probabilities computed for the bin mean, *v* 4. In the table, the three differences are indicated for each variable by the symbols *LT-BL* (left minus baseline), *RT-BL* (right minus baseline), and *RT-LT* (right minus left). The *t*-scores are written with a minus sign if the corresponding quantity had a negative displacement. The correlation coefficients between the difference pairs for each of the 7 variables and the difference pairs for *v* 4 are also indicated.

It is natural for the displacement probabilities for variables *v* 4 and *v* 5 to be similar, since these variables have a correlation of about .71 for each of the three differences. Thus *v* 4 has probabilities of .000076 and .00005 for *LT-BL* and *RT-LT*, and *v* 5 has similar probabilities for these differences. These small probabilities represent the main anomalous effect. However, if we look at variables *v* 6, *v* 7, *v* 8, *v* 10, *v* 11, and v 14, we can see that they also tend to have moderately low probabilities for *LT-BL* and *RT-LT*, even though they have very small correlations with *v* 4.

One should ask whether or not this might be statistically significant. The answer is that, individually, the displacements of these variables are not highly significant, but they *are* highly significant when taken together as a group. Since the variables tend to be mutually uncorrelated, this can be shown by summing them up and examining the displacement of the sum.

With this aim in mind, the calculations of Table IV were also performed for two composite variables:

$$v\ 15 = Z(v\ 6) + Z(v\ 7) + Z(v\ 8) - Z(v\ 10) - Z(v\ 11) + Z(v\ 14) \quad (8)$$

and

$$v\ 16 = Z(v\ 6) + Z(v\ 7) + Z(v\ 8) + Z(v\ 14) \quad (9)$$

where *Z* (variable) is that variable shifted so as to have mean 0 and scaled so as to have a standard deviation of 1 over the 1510 tripolar sets. The transformation *Z* was applied so as to make the magnitudes of the variables comparable, and thereby prevent the high magnitude variables in the sum from overshadowing the contributions of the low magnitude variables. In *v* 15 the variables *v* 10 and *v* 11 were given a minus sign due to the fact that they vary in a direction opposite to that of variables *v* 6, *v* 7, *v* 8, and *v* 14. (This can be seen by examining Table IV.)

The same computations were performed for *v* 15 and *v* 16 that were performed for *v* 4, *v* 5, *v* 6, *v* 7, *v* 8, *v* 10, *v* 11, and *v* 14, and the results are listed in Table IV. These results indicate that the mild effects noted for the constituent variables of *v* 15 and *v* 16 do seem to add up. We see from the table that *v* 15 has probabilities of .00015 and .00075 for *LT-BL* and *RT-LT*, and *v* 16 has corresponding probabilities of .00057 and .00034. These probabilities indicate that the group behavior of the variables making up *v* 15 and *v* 16 is significantly different from what would be expected by chance.

Here the objection could be raised that perhaps we have tried our analysis with many different variables, and have presented the results for *v* 15 and *v* 16 because they turned out to give us low probability values. These values would then be of no real significance, because one can always find a spurious effect that looks significant if one juggles the data sufficiently.

The answer to this objection is that we did not do this. The variables in Table I were selected initially on the basis of our understanding of the RMC, and no other variables of this type have been considered. The subset of these 14 variables listed in Table III was chosen solely on the basis of the correlation coefficients listed in Table II, the object being to find a subset of *v* 1, ... , *v* 14 with mutual correlations of the smallest possible magnitudes.

We can see from Table IV that all 6 of the variables of Table III show a noticeable tendency towards systematic drift in the cases *LT-BL* and *RT-LT*. The two sums, *v* 15 and *v* 16, show that these tendencies, taken together, are statistically significant. The composite variables *v* 15 and *v* 16 are the only ones considered; they were not chosen from a larger set of composite variables.

We also performed the calculations of Table IV for the full collection of 1510 tripolar sets of runs. In this case it turned out that *v* 15 has probabilities of .0079 and .0081 for *LT-BL* and *RT-LT*, and *v* 16 has probabilities of .024 and .0034. In general, one tends to obtain lower probabilities for the 1131 case subset than for the full data set of 1510 cases. What is happening here is that anomalous effects are much weaker in the part of the data set involving either multiple operators or operators who could not see the RMC in operation.

We note that the correlation coefficient of *v* 4 and *v* 15 is .073 (over the entire data set), and that of *v* 4 and *v* 16 is .028. The correlation coefficients between the difference pairs for *v* 15 and *v* 16 and the corresponding difference pairs for *v* 4 are listed in Table IV, and also lie between 0 and .1. (Recall that the difference pairs are *v* 15 (*i*, *LT*) - *v* 15 (i, *BL*), and so on.) Although these correlation coefficients are small, they are significantly different from 0 in the positive direction. This can be seen by applying the formula,

$$t = r \left[(N - 2) / (1 - r^2) \right]^{1/2} \quad (10)$$

where N is the number of cases (1510 for the whole data set and 1131 for the subset), r is the correlation coefficient, and t is interpreted as a t value.[8] For the correlation coefficients in question we obtain t values ranging from 1.09 to 2.97.

The improbable displacements of v 15 and v 16 cannot be accounted for by the hypothesis that v 15 and v 16 shift in correlation with the shift in v 4. In fact, the displacements of v 15 and v 16 for *LT-BL* and *RT-LT* go in the direction opposite to the corresponding displacements of v 4. These displacements constitute an anomalous effect that is independent of the original anomalous effect reported by Jahn and his colleagues. Moreover, this effect is due to the behavior of variables (the constituents of v 15 and v 16), which could not be observed by the operators and were not the objects of operator intentions.

We have examined the 4530 pairs, (v 4, v 15), we find that their distribution can be approximated by a bivariate normal distribution. The same is true of the pairs, (v 4, v 16). This means that Equation (5) should apply with $v = v\ 4$ and $u = v\ 15$ or v 16. This equation implies, for example, that

$$[E(v\,15 \mid LT) - E(v\,15 \mid BL)] / s_{v15} = r\,[E(v\,4 \mid LT) - E(v\,4 \mid BL)] / s_{v4} \quad (11)$$

where r lies between 0 and .1. But this is contradicted by the actual behavior of v 15, which shifts strongly in the positive direction in this case, even though v 4 shifts strongly in the negative direction.

In addition to the calculations presented in Table IV, we produced graphs showing the behavior of the 10 variables, v 4, ... ,v 16. For these graphs we adopted the following procedure. First, the variables were all scaled so as to have mean 0 and standard deviation 1. That is, we replaced vk by $Z(vk)$ for each of the variables. We then formed all of the pairs, (v 4, v k), for k = 5, 6, 7, 8, 10, 11, 14, 15, 16. For the pair, (v 4, v k), define the two dimensional random variable (X_{2n-1}, Y_{2n-1}) to be (v 4, v k), computed for the nth tripolar set with leftward intention. Similarly, define (X_{2n}, Y_{2n}) to be ($-v$ 4, $-v$ k), computed for the nth tripolar set with baseline intention. We plotted the random walk generated by steps of (X_n, Y_n) for $n = 1, \ldots, 2 \times 1510$.

We will let (S_n, T_n) represent the random walk generated by steps of (X_n, Y_n). Thus, $S_o = T_o = 0$ and

$$(S_n, T_n) = (X_n, Y_n) + (S_{n-1}, T_{n-1}) \quad (12)$$

For each tripolar set of runs, this random walk takes one step of (v 4, v k) for the run of intention, *LT*, and one oppositely directed step of (-v 4, -v k) for the run of intention, *BL*. This gives us an idea of the comparative behavior of v 4 and v k in the case *LT-BL*. Similar random walks were plotted for the cases *RT-BL* and *RT-LT*. We also plotted these random walks for the subset of 1131 tripolar sets in which a single operator was watching the RMC for the subset of tripolar sets generated by operator 10, and also for the subset in which operator 10 was watching the RMC.

For a given angle, Φ, the random walk can be projected onto the line through the origin at angle Φ. This projected random walk is

$$U_n = cos\ (\ \Phi\ )\ S_n + sin\ (\ \Phi\ )\ T_n \tag{13}$$

The standard deviation of U_n is $[\ n\ (1\ +\ sin\ (\ 2\Phi\ )\ r\ ]^{1/2}$, where r is the correlation coefficient for X_n and Y_n. If we assume that v 4 and v k keep the same statistical properties for tripolar sets in the interval from $n = 1$ to 2×1510, and that operator intention is irrelevant to their behavior, then r is the correlation coefficient between v 4 and v k, as estimated in Table III. (This table should be supplemented with values for r of .073, .028, .814 for (v 4, v 15), (v 4, v 16), and (v 15, v 16), respectively.) Although there is some change in the statistical properties of the vk's with the passage of time (represented by n), the curve,

$$R = 2\ [\ N\ (\ 1\ +\ sin\ (\ 2\Phi\ )\ r\ )]^{1/2} \tag{14}$$

provides a two standard deviation limit that can be used to evaluate the behavior of (S_n,T_n,) for $n = 1, \ldots, N$.

Figures 3(a–e), 4, and 5 (a-c) show some of the plots of these random walks, along with their estimated two standard deviation limits. In each graph, S_n is plotted along the x-axis and T_n is plotted on the y-axis. Figure 3(a) shows the behavior of v 4 and v 16 for the subset of the 1510 tripolar sets generated by operator 10, the most successful operator in the RMC experiments. This figure shows the case of *LT-BL*, or left minus baseline. In this figure we see that the random walk moves to the left and up in nearly a straight line. The leftward movement is the anomalous displacement of the bin mean, v 4, and the upward movement is the anomalous displacement of v 16. Figures 3(b-e) show the corresponding behavior of v 4 and vk, where $k = 6, 7, 8$, and 14. These are the variables that sum up to form v 16 (see Equation 9).

Figure 3(a) should be compared with Figure 4, which shows the (v 4, v 16) random walk in the case of *LT-BL* for the subset of runs in which operator 10 watched the RMC in action. We note that the random walk in Figure 4 moves further out from the two standard deviation limit than does the random walk in Figure 3(a); this is due to the fact that the anomalous effects seem to occur

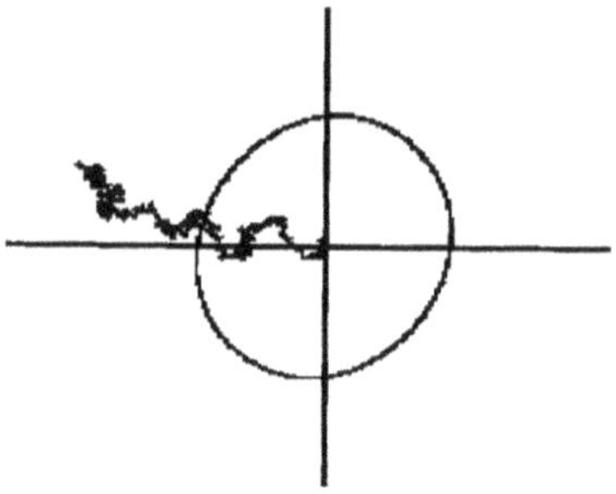

Figure 3(a). Operator 10, left-baseline. Parameters (4,6) are the (x,y) axes.

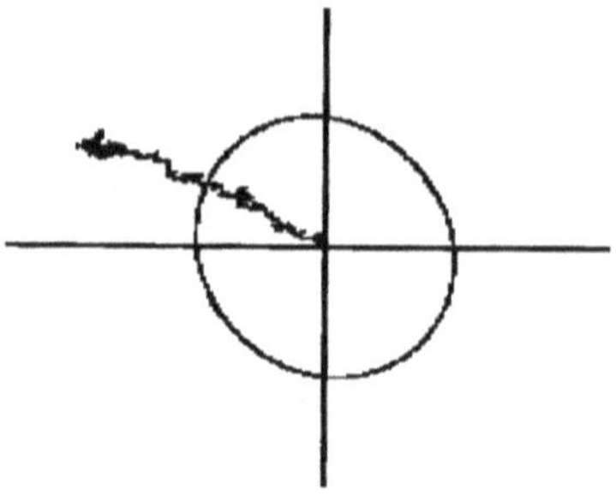

Figure 3(b). Operator 10, left-baseline. Parameters (4,7) are the (x,y) axes.

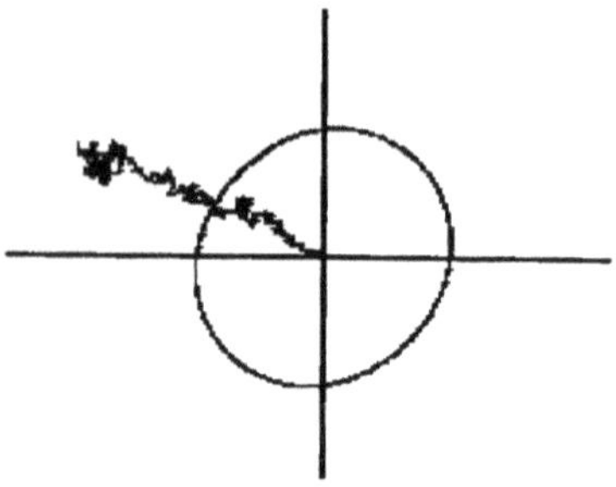

Figure 3(c). Operator 10, left-baseline. Parameters (4,8) are the (x,y) axes.

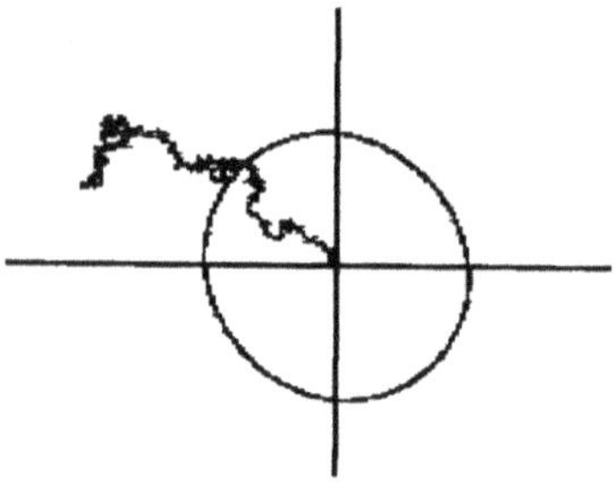

Figure 3(d). Operator 10, left-baseline. Parameters (4,14) are the (x,) axes.

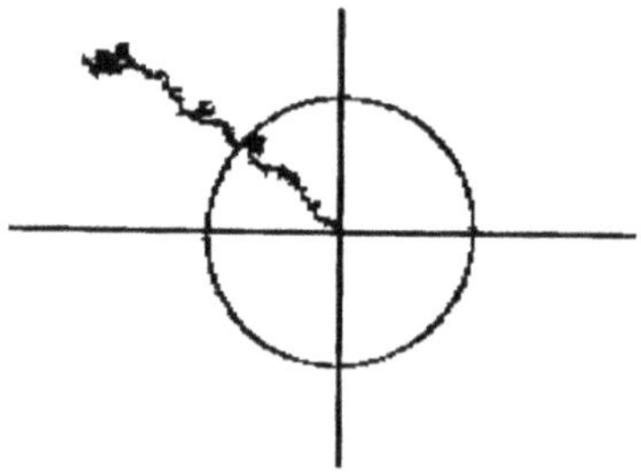

Figure 3(e). Operator 10, left-baseline. Parameters (4,16) are the (x,y) axes.

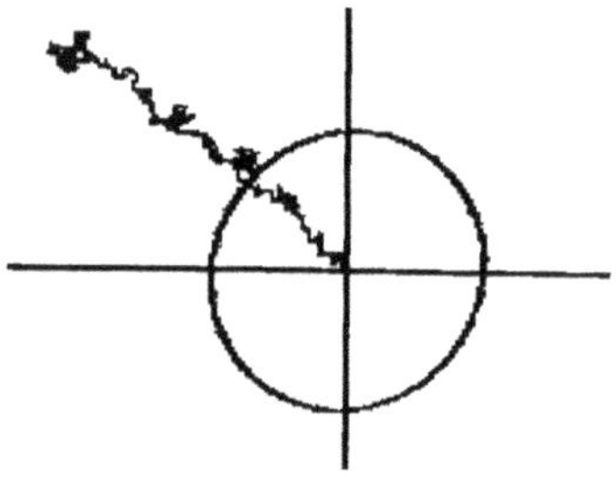

Figure 4. Operator 10, left-baseline. Parameters (4,16) are the (x,y) axes. Case of 1131 runs.

predominantly in the runs involving a single operator who is able to observe the RMC.

Figures 5(a-c) show the behavior of *v* 4 and *v* 16 for all 1510 tripolar sets. In this figure, pans (a), (b), and (c) are for the respective cases of *LT-BL*, *RT-BL*, and *RT-LT*. Figure 5(b) shows that in the *RT-BL* case, *v* 4 barely moves from the origin, while *v* 16 shows considerable activity. This rather puzzling nonrandom behavior occurs in the *RT-BL* cases for practically all of the variables.

To test whether or not this apparent nonrandomness is illusory, we performed a number of probabilistic experiments using the RMC data. The results of the first of these experiments are summed up in Figure 6. There we have superimposed 100 random walks, which are generated in nearly the same way as the random walks in the case of *v* 4 *vs*. *v* 16 for all tripolar sets. The only difference is that in each tripolar set, the + (*v* 4, *v* 16) and - (*v* 4, *v* 16) steps of the random walk are chosen at random from the three runs in that set, instead of being chosen on the basis of intention. The random choices were made using a pseudo-random number generator. The idea here is to create random walks that are identical to the random walks of Figures 5(a–c), with the exception that information concerning operator intentions is erased by random scrambling. Figure 6 shows that these random walks show no particular tendency

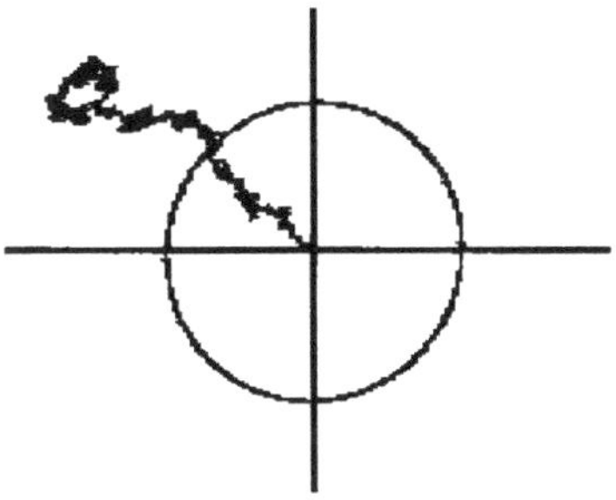

Figure 5(a). All operators, left-baseline. Parameters (4,16) are the (x,y) axes.

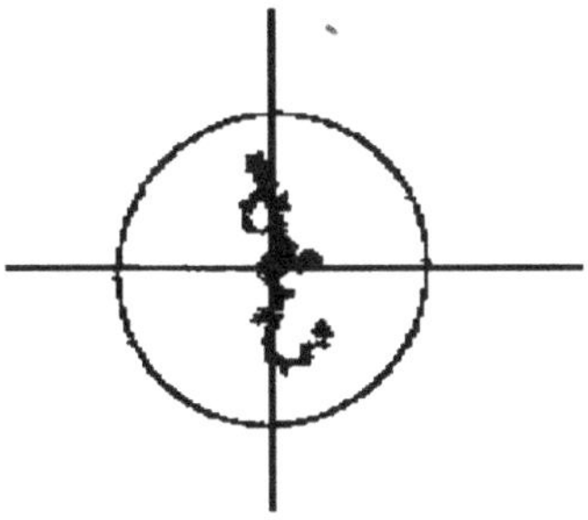

Figure 5(a). All operators, right-baseline. Parameters (4,16) are the (x,y) axes.

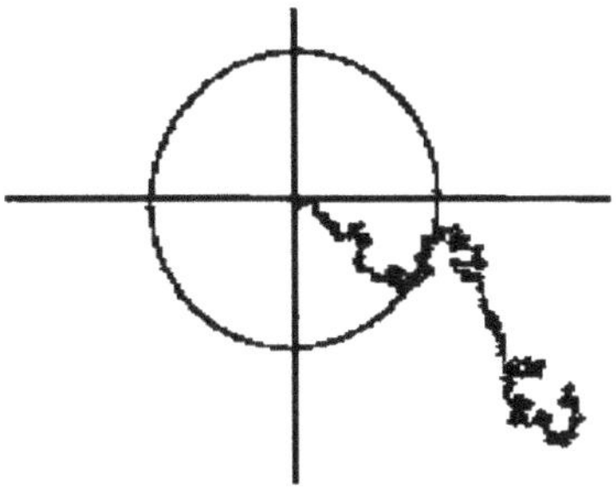

Figure 5(c). All operators, right-left. Parameters (4,16) are the (x,y) axes.

for systematic drift, apart from a slight tilt with positive slope which may be due to the small positive correlation between v 4 and v 16. Thus, all of the anomalous effects in Figures 5(a–c) seem to depend on the information regarding operator intentions.

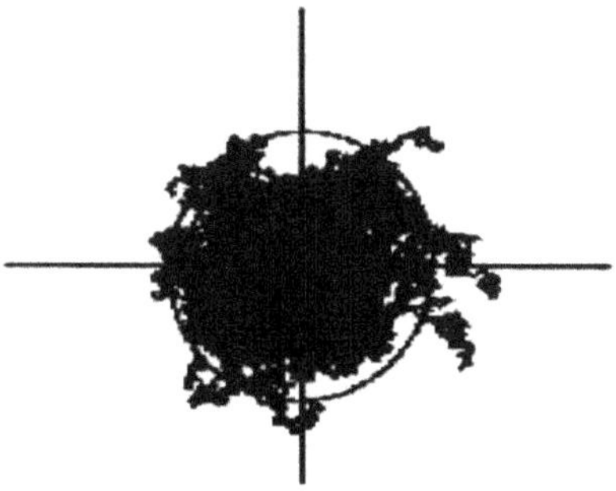

Figure 6. Superposition of 100 RMC random walks with randomized intention. Parameters (4,16) are the (x, y) axes.

The results of the second experiment are shown in Figure 7. This experiment is the same as the first, with one exception: In the random walks, if a step is positive in v 4, then it is rejected with a certain probability; if it is negative then it is always accepted. This rule has the effect of imposing an artificial bias against movement in the positive v4 direction (the positive x-axis in the plots). This is exactly the kind of bias we would postulate in the "data selection" theory described in the previous section. If that theory is correct, then these random walks should be similar to the one plotted in Figure 5(a).

We can see from Figure 7 that, with a couple of exceptions, there is no tendency for the random walks generated in this way to mimic Figure 5(a), the real random walk for variables v 4 and v 16 and intentions *LT-BL*. The artificially generated random walks do show a tendency to drift systematically to the left, but instead of also drifting in the positive y direction, they exhibit, if anything, the opposite tendency.

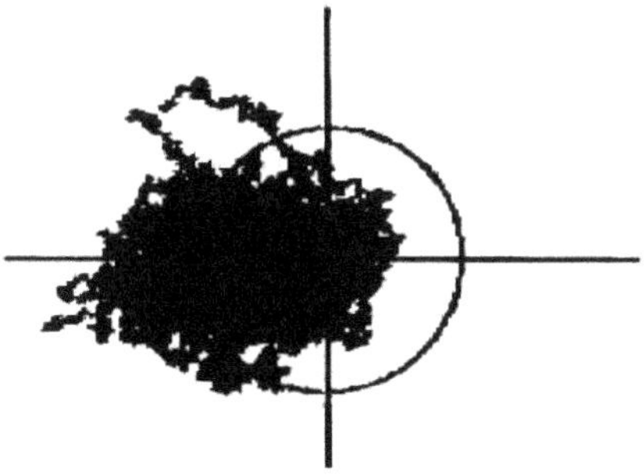

Figure 7. Superposition of 100 RMC random walks with randomized intention and artificial selection of parameter 4 low (ie. to the left). Parameters (4,16) are the (x,y) axes.

We also considered the following hypothesis: Suppose there is a variable, w, that is weakly correlated with both v 4 and v 16. Perhaps the joint displacements of v 4 and v 16 can be explained as a result of selection applied to w. To investigate this hypothesis properly, one would have to consider many different candidates for *w*. We investigated only two possible *w*'s, and neither one could account for the observed displacements of *v* 4 and *v* 16.

4. Conclusion

The basic conclusion that emerges from this study is that in addition to the original anomalies reported by Jahn and his colleagues, the statistical behavior of the bin distributions seems to display systematic patterns that correlate with human intentions, but are not related in an obvious way to the conscious content of those intentions. Thus, the data show a systematic shift in *v* 4, the bin mean, but they also show shifts in *v* 15, and *v* 16. These, in turn, are due to the summing up of shifts in the constituent variables of *v* 15 and *v* 16, namely *v* 6, *v* 7, *v* 8, *v* 10, *v* 11, and *v* 14. The shifts in *v* 15 and *v* 16 are somewhat less improbable than the shift in the bin mean, but they are nonetheless statistically significant. This is especially true for the set of 1131 runs in which a single operator was present during the running of the RMC (see Table IV).

The shifts in *v* 15, *v* 16, and their constituent variables cannot be accounted for by intervariable correlations with the bin mean that could be due to the physical characteristics of the RMC device. The reason for this is that we can calculate the correlations between *v* 4 and these two variables over the whole data set, without taking operator intentions into account. When this is done, it is found that the resulting correlations go in the wrong direction to account for the observed shifts in *v* 15 and *v* 16.

These findings tend to rule out a class of theories, called selection theories, which might be invoked to explain the anomalies in the RMC data. In particular, they tend to rule out the possibility that the main anomalous effect in the bin mean was obtained by conscious or unconscious data selection by the experimenters. Certainly a drift in *v* 4 in a given direction can be obtained by systematically throwing out a certain percentage of the runs which do not tend in that direction. However, as we have seen, such editing of the data will not generate the additional shifts that we have observed in certain other variables, and it seems unlikely that these variables could have been the object of additional data selection by the experimenters.

Theoretical calculations suggest that it would be difficult for the experimenters to generate the observed anomalous effects by exerting small, normal forces on the RMC (of the kind that could be produced by a human body at a distance from the machine). Forces of the strength that could be produced by an observer watching the RMC *could* influence the RMC in accordance with the observer's intention, if these forces could be generated with very precisely specified magnitudes. But to do this, it would be necessary for the observer to somehow unconsciously determine what these magnitudes should be – a task which would be difficult to carry out even with a supercomputer and an accurate mathematical model of the RMC.

What we are left with is the conclusion that some unknown agency affects the behavior of the RMC in accordance with conscious intentions of the operators. This agency not only affects the bin mean, which is the object of the operators' intentions, but it also affects other aspects of RMC behavior which would be expected to vary independently of this variable. To learn more about what is happening here, we would recommend future experiments in which many different aspects of a physical process are monitored, while efforts are being made to influence particular features of that process. Such experiments would reveal whether or not the phenomena we have observed here generally occur, and they may give further insight into the causes of these phenomena.

References and Notes

1. R. D. Nelson, B. J. Dunne, and R. G. Jahn, *Operator Related Anomalies in a Random Mechanical Cascade Experiment*, (School of Engineering/Applied Science, Princeton University, Princeton, NJ, 1988).
2. Nelson *et al.*, 4–5.
3. Nelson *et al.*, 33–34.
4. P. Cvitanovich, *Universality in Chaos* (Bristol, UK: Adam Hilger, 1984).
5. Cvitanovich, 92
6. E. P. Wigner, "Physics and the Explanation of Life," *Foundation of Physics* 1, 35 (1970).
7. Nelson *et al.*, 8.
8. G. W. Snedecor and W. G. Cochran, *Statistical Methods* (Ames, IA: Iowa State University Press, 1980), 185.

CHAPTER 4

Emperor Aśoka and the Five Greek Kings*

Produced by:
Bhaktivedanta Institute, Alachua, FL (1994)

Note: The citations that appear in this essay are in a draft form. References to images "Figure 1," "Figure 2," and "Appendix 3" remain as placeholders, since the contents that would be identified with them did not appear with the original manuscript.

In the nineteenth and early twentieth centuries, light was shed on the ancient history of India by the discovery and decipherment of a large number of royal edicts carved in forgotten alphabets on rocks and pillars. The edicts heralded the achievements of a king named Priyadarśī in "moral conquest" or *dharma-vijaya* – an ambitious program of public works and state-controlled moral reform for which he claimed success at home and in many foreign territories.

Since Priyadarśī's edicts were found over a broad area of the Indian sub-continent, ranging from northern Pakistan to South India, it appeared that he was a powerful emperor of great historical importance. At first it was difficult to identify him with any known historical figure. But scholars surmised that Priyadarśī might be Aśoka, an emperor who is mentioned in the dynastic lists of the *Purāṇas* and who is glorified in the Sri Lankan Buddhist text *Mahāvaṁsa* for his efforts to spread Buddhism. They therefore began to refer to Priyadarśī's inscriptions as the edicts of Aśoka. They believed this identification was clinched by the discovery of inscriptions at Maski in 1919 and Gujarra in 1954 that referred to Priyadarśī as Aśoka.[1]

* Version with attempted footnote numbers.

1 Woolner, 1924, p. xx and Sen, 1956, pp. 10, 51.

Most of the Aśokan edicts were written in various dialects of Prakrit, an ancient Indian language closely related to Sanskrit. Many were written in the Brāhmī alphabet, which is the ancestor of many Indian alphabets of today, and a few were written in Kharoṣṭhī, an alphabet related to the Aramaic script of Persia and the Near East.[2]

In 1838, James Prinsep reported the successful decipherment of the Brāhmī alphabet, and he published the first translation of an Aśokan edict. He also reported the translation of the Aśokan rock edict 2, which had been found and transcribed at Girnār in the province of Gujarat and at Dhauli in Orissa.[3]

This achievement was accompanied by a remarkable discovery. Prinsep read what he took to be the name of Antiochus the Great in the Girnār and Dhauli inscriptions. Antiochus the Great was the king of Syria in 223–187 B.C., and he ruled a war-torn empire extending from Asia Minor in the West to Persia in the East. If Antiochus was truly mentioned by Aśoka, this would fix the date of Aśoka's reign and shed light on the political relations between India and the West in ancient times.

Later in the same year, Prinsep reported another important discovery that he said was "most attractive to all who have been nurtured in the school of western classical associations."[4] This time he read in Girnār rock edict 13 the names of three additional Hellenistic kings: Ptolemy, Magas, and possibly Antigonus. There was only one candidate for Magas: king Magas of Cyrene who ruled in 300–258 B.C. To make the other kings contemporaries of Magas of Cyrene, Prinsep argued that Ptolemy should be Ptolemy Philadelphus of Egypt and Antiochus should be one of the predecessors of Antiochus the Great, either Antiochus Soter or Theos. For Antigonus he suggested king Antigonus Gonatas of Macedonia.[5]

A few years later in 1845, another name was added to the list. E. Norris deciphered the Aśokan inscriptions of Shāhbāzgarhi in northern Pakistan. Like the Girnār and Dhauli inscriptions, these were divided into 14 texts that are called rock edicts since they are inscribed on natural rock surfaces. The rock edicts at Shāhbāzgarhi and other sites proved to contain essentially the same material as those at Girnār, and in rock edict 13 Norris found the four names

2 Beginning in 1914, Aśokan edicts written in the Aramaic language were found in the area of Afghanistan. Edicts written in Greek began to turn up in this area in 1957 (Mukherjee, 1984)

3 Prinsep, 1838a

4 Prinsep, 1938b, p. 219

5 Prinsep, 1838b, pp. 225–26

already discovered. Norris also added a fifth – an Alexander whom he tentatively identified as king Alexander of Epirus.[6]

Scholars quickly became convinced that the five names discovered in the Aśokan inscriptions had been properly identified as Hellenistic kings living in the period between Alexander the Great and the extension of Roman power to Asia. In books on ancient Indian history, the identifications of the five names are typically presented as follows:

Aśokan name	**Hellenistic king**	**Reign (years B.C.)**
Aṁtiyoka	Antiochus Theos of Syria	261–246
Turamāya	Ptolemy Philadelphus of Egypt	285–247
Aṁtikini	Antigonus G[7]onatas of Macedonia	276–239
Maga	Magas of Cyrene	c. 300–c. 250
Alikasuṁdara	Alexander of Epirus	272–c. 258
	or Alexander of Corinth	252–c. 244

Table 1. Accepted identifications of the five kings of Aśoka's rock edict 13.[8]

Scholars date Aśoka's reign on the basis of his presumed mention of the five Hellenistic kings. If we examine the regnal periods in Table 1, we can see that all five kings were reigning simultaneously in the brief period from 261 B.C. to 258 B.C. This period should be regarded as approximate since the dates assigned to these kings vary somewhat from one textbook to another.

Many scholars maintain that rock edict 13 must have been issued during this period since it seems to refer to all five kings as contemporary rulers. The rock edicts are said to indicate that they were compiled 12 years after the coronation of Priyadarśī (Bongard-Levin, 1985, p. 89). If we add 12 to 261 B.C. we get 273 B.C. for the date of the coronation. The *Purāṇas* say that Aśoka reigned for 36 years, and the Sri Lankan chronicles give a period of 37 years. Subtracting these figures from 273 B.C., we get 236–237 B.C. for the end of Aśoka's reign.

Of course, there are many different ways to derive dates for Aśoka's reign from this evidence. Nikam (1966, p. 1) and Majumdar (1960a, pp. 108, 115) give a period of about 273 B.C. to 232 B.C. Bongard-Levin (1985, p. 90) gives 268 or 265 B.C. to roughly 232–228 B.C. The essence of the argument is that

6 Norris, 1846

7 xxxx

8 Bhandarkar, 1955, p. 43

if the five kings have been properly identified and Aśoka was their contemporary, then Aśoka's reign must overlap with all of their reigns. Then again, if the five kings have been properly identified and Aśoka merely heard about them, it follows that the end of his reign must come after the beginnings of all their reigns. In either case, Aśoka cannot antedate any of these kings.

But have the five kings been properly identified? How do we decide whether a similarity between two names is due to historical identity or is just coincidental? Clearly, the answers to these questions do not depend simply on the names themselves. To arrive at a satisfactory answer, we must examine the names in their historical setting. I will begin by asking how well Priyadarśī's program of *dharma-vijaya* fits into the Hellenistic societies of the 3rd century B.C.

I. AŚOKA'S MEDITERRANEAN MISSION

If Aṁtiyoka and his colleagues were really the Hellenistic kings of Table 1, then Priyadarśī was apparently running a large scale propaganda and foreign aid program in the eastern Mediterranean region. At least, this is the impression conveyed by the two Aśokan edicts that mention the famous names. The first of these is rock edict 2:

> "Everywhere in the dominions of King Priyadarśī, as well as the border territories of the Choḷas, the Pāṇḍyas, the Satiyaputra, the Keralaputra [all in the southern tip of the Indian peninsula], the Ceylonese, the Yona king named Aṁtiyoka, and those kings who are neighbors of Aṁtiyoka – everywhere provision has been made for two kinds of medical treatment, treatment for men and for animals.
>
> "Medicinal herbs, suitable for men and animals, have been imported and planted wherever they were not previously available. Also, where roots and fruits were lacking, they have been imported and planted.
>
> "Wells have been dug and trees planted along the roads for the use of men and animals" (Nikam, 1966, p. 64).

Here the translator gave the name of the Yona king as "Antiochos," but I have restored it to "Aṁtiyoka" in accordance with the actual text of the edict. The second edict mentioning the names is rock edict 13. This edict mentions the names of Aṁtiyoka's four colleagues, and it also makes strong claims regarding the ideological impact of Aśoka's program:

> "King Priyadarśī considers moral conquest [*Dharma-vijaya*] the

> most important conquest. He has achieved this moral conquest repeatedly both here and among the peoples living beyond the borders of his kingdom, even as far away as 600 *yojanas*, where the Yona king Antiyoka rules, and even beyond Antiyoka in the realms of the four kings named Turamaya, Antikini, Maka, and Alikasudara, and to the south among the Cholas and Pāṇḍyas as far as Ceylon" (Nikam, 1966, p. 29).

The program outlined in these edicts is quite realistic for India. The moral laws that Priyadarśī was advocating are part of India's ancient tradition of *sanātana-dharma*, and Indian kings were expected to uphold these principles. The fact that the principles of *dharma* were once prominent throughout the Indian subcontinent suggests that *dharma-vijaya* was accomplished by some Indian rulers, and these may well have included Priyadarśī.

It was also traditional for Indian kings to engage in the public works mentioned in edict 2, and it would have been realistic for Priyadarśī to arrange for such works within India, even in areas not under his direct authority. In India such public works are well known. For example, J. Wilson, who recorded the Girnār edicts, made the following comments about the Jains in Gujarat: "They maintain *pinjarāpurs*, or brute hospitals, like the Banyas of Surat, in many of the towns both of the peninsula and province of *Gujerāt*; and practice to a great extent the long forgotten, but now restored, edict of Asoka" (Prinsep, 1838c, p. 337). Likewise, Perry (1851, p. 152) mentions the universal appearance of *dharmasālas* in all Hindu states. He also refers to groves of artificially planted mangos and other fruit trees that sometimes extend for miles.

But was Priyadarśī's program realistic for the eastern Mediterranean region, where philosophies, customs, and state policies were greatly different from those of India? Although this seems doubtful, Prinsep thought so. Regarding Egypt, he said,

> "We can easily believe that its enlightened sovereign would afford every encouragement to the resort of Indians thither, for the sake of promoting that commerce with India which was so fertile a source of enrichment: and indeed history tells us that Ptolemy Philadelphus deputed a learned man named Dionysius to India to examine the principal marts on the western coast, and in the interior" (Prinsep, 1838b, p. 226).

He also suggested that Ptolemy would have been eager to study the philosophy of the Indian *brachmani* and *sramani*, and he expected that much evidence would be found of the influence of Buddhistic principles on the "prevailing opinions of the day" in Antioch and Alexandria (Prinsep, 1838b, pp. 226–27).

But it turns out that historians have not uncovered any reference to Priyadarśī or his Mediterranean mission in the histories of the Hellenistic kingdoms. The surviving Greco-Roman records do not name the Indian king to whom Dionysius was sent (Bevan, 1968, p. 155). They mention only two Indian kings known to the successors of Alexander the Great, and neither of these is a good candidate for Aśoka (Majumdar, 1960b). One is Amitrochades, who scholars identify as Aśoka's father. The other is Sophagasenus, a minor king who had some dealings with Antiochus the Great. The absence of recognizable Greco-Roman references to Aśoka stands out as an anomaly in the standard reconstruction of Indian history.

If we examine Priyadarśī's edicts, we find that most of them describe a non-denominational social program involving public works and moral instructions. In contrast, the Ceylonese Buddhist chronicle *Mahāvaṁsa* says that Aśoka sent out missionaries who explicitly preached Buddhist doctrines, built Buddhist temples, and converted large numbers of people to Buddhism (Upham, 1833, pp. 76–83). This discrepancy suggests that the Aśoka of the *Mahāvaṁsa* and the Priyadarśī of the edicts may not have been the same person, even though the Maski and Gujarra edicts seem to suggest that they were the same.

If they were not the same person, then the edicts of Priyadarśī tell us nothing about Aśoka. If they were the same, then we would expect to see signs of explicit Buddhist influence in the Hellenistic kingdoms of the 3rd century B.C. As we have seen, Prinsep certainly expected this. Yet the historian R. C. Majumdar said, "Greece knew nothing of Buddhism previous to the rise of Alexandria in the Christian era. Buddha was first mentioned by Clement of Alexandria (A.D. 150–218)" (Majumdar, 1951, p. 616). The absence of signs of Buddhist influence in the Hellenistic world is strong evidence suggesting that the agents of Aśoka-Priyadarśī were not active there.

Another reason for doubting the reality of Priyadarśī's Mediterranean mission lies in the list of Hellenistic kingdoms in Table 1. Egypt, Syria, Macedonia, Cyrene, Epirus, and Corinth are a very heterogeneous group. They range in size from Syria, which controlled territories from Asia Minor to the borders of Aśoka's kingdom, to Cyrene and Epirus, which were small principalities dominated by stronger neighbors (Egypt and Macedonia). Corinth, of course, was a Greek city-state.

Why would Priyadarśī publicize programs in these particular places and not mention other states in the same general area? Here is a partial list of additional states (or confederations) that were existing during the general time period assigned to Aśoka's reign. Some of these states were initially provinces or Satrapies of Alexander's empire, and they acquired independence by revolting against his successors.

State	Region	Kings	Key Dates (B.C.)
Achaean League	Greece		c. 280
Aetolian League	Greece		279–217
Athens	Greece		free c. 262
Bactria	N.W. of India	Diodotus I & II	256–235
Bithynia	Asia Minor	Nicomedes I	279–225
Parthia	N. Iran	Arsaces I	c. 250
Pergamum	Asia Minor	Philataerus	282–263
		Eumenes	263–241
Pontus	Asia Minor	Mithradates	280

Table 2. Some states existing during the reign assigned to Aśoka (Davis, 1973, and Kinder, 1974).

Bactria, in particular, would have been situated on Aśoka's western frontier. This territory revolted against Antiochus Theos of Syria in about 256 B.C. under the leadership of Diodotus I, and it quickly became a major political power on the western border of India. Before 256 B.C., Bactria would have been a Satrapy of Antiochus. But even then it was a logical target for Priyadarśī's program of propaganda and foreign aid. Yet he does not seem to mention Bactria or any of the other countries listed in Table 2.

Modern scholars have taken a great deal of information about Aśoka from Sri Lankan Buddhist chronicles, such as the *Mahāvaṁsa*. We might therefore ask what these sources have to say about Aśoka's mission to the West. The *Mahāvaṁsa* states that Aśoka's chief priest Moggaly-Tisse-Maha sent out nine senior priests as Buddhist missionaries. These are named in Table 3, along with the countries to which they were sent (Upham, 1833, pp. 76–83).

All of the regions mentioned in this table are in India or directly adjacent to India. Why is there no mention of Egypt, Syria, and Macedonia in this list? According to historian Vincent Smith, "The exclusion of the Hellenistic kingdoms from the Ceylon list is easily explained when we remember that these kingdoms had ceased to exist centuries before that list was completed" (Smith, 1964, p. 44).

Majumdar (1960a, p. 184) held that the *Mahāvaṁsa* was probably written by the poet Mahānāma in the late 5th century A.D. If this poet could not describe the history of long-vanished kingdoms, then how was he able to write about the long-vanished Aśoka? The story of Aśoka's successful preaching in the Mediterranean countries should have been treasured by the Buddhist chroniclers as one of the great victories of their faith. Even if Aśoka made the whole thing up, it still made a great story – and the *Mahāvaṁsa* is certainly not lacking in amazing stories. The testimony of the *Mahāvaṁsa* clearly tends to support the idea that Aśoka-Priyadarśī was not in contact with the heirs of Alexander the Great.

Preacher	Region
Matjantica	Cāsmira and Gandāre (Kashmir and Kandahar)
Mahadewe	Mahimandelle (Mysore)
Racsita	Wannewahse (North Kannara)
Yoneke-Darmeracsite	Aperanta (coast north of Bombay)
Mahadarmeracsite	Rawstra (West Central India)
Maharacsita	Yonacca (N.W. frontier provinces)
Matjeoma Maher	Hemmewanta (Himalayan region)
Seeneca	Swarnewarna (in Burma)
Mihidu and others	Sri Lanka

Table 3. Buddhist missionaries sent out by Aśoka's chief priest and the lands where they preached.

Some commentators have, indeed, denied the reality of Priyadarśī's Western mission and dismissed his statements about it as "oriental vanity," or mere idle boasting (Perry, 1851, p. 167). The prominent indologist H. H. Wilson even went so far as to deny that the Priyadarśī of the edicts existed as a historical personality. He suggested that "the rulers of several countries or influential religious persons adopted the shadow of a name, to give authority to the promulgation of edicts intended to reform the immoral practices of the people" (Wilson, 1850, p. 250). According to Wilson, the names Antiochus, Ptolemy, Antigonus, Magas, and Alexander had become known in India, and they were simply used for prestige by the edict writers. If Wilson was right, references to Hellenistic kings in the edicts would tell us nothing about the date of Aśoka.

But suppose Priyadarśī was real and was identical with the Aśoka of the

Purāṇas. The statements in the Aśokan edicts sound eminently practical and realistic when applied to India or bordering territories under strong Indian influence. Aśoka-Priyadarśī was evidently not given to idle boasting at home. So why should he make wild claims about nonexistent missionary activities in distant countries?

Yet if his western programs really took place, why did they leave no trace in the Western world? Whether they took place or not, why were his claimed successes forgotten by the Buddhist chroniclers who glorified him as a great propagator of Buddhism? The standard theory is not consistent with the historical evidence.

One solution to this problem is to suppose that Priyadarśī's mission was limited to the immediate vicinity of India and that Aṁtiyoka and his colleagues were minor kings of small Indian states. This would explain why the Buddhist chronicles do not name them and why there is no Western reference to Aśoka. It also makes Aśoka's claims uniformly realistic and consistent.

This proposal also accounts for the fact that Aṁtiyoka is referred to in the edicts as Yona-rāja – a rather humble title. Prinsep (1838c, pp. 346–48) pointed out that a Sanskrit inscription near Girnā mentioned a Yavana-rāja named Tushaspa who was under the orders of Aśoka Maurya. Prinsep translated Yavana-rāja as "Greek officer," and, interestingly enough, he pointed out that Tushaspa is a Persian name. The Prakrit word Yona is generally taken to be synonymous with the Sanskrit word Yavana. This suggests that Yona-rāja in Aśoka's edicts might have designated an inferior king or officer. It also suggests that such a person was not necessarily Greek. He might have been a Persian or perhaps a member of some other non-Indian ethnic group.

Antiochus Theos was a successor of Alexander the Great and the Persian King-of-Kings. One would think that Priyadarśī, as an expert diplomat, would refer to him accordingly. Even if Priyadarśī was making up a story about some famous Western kings, the story sounds better if the objects of successful preaching are important rulers with impressive titles. In contrast, the term Yona-rāja might be appropriate for a minor ruler of some outcast group on the fringes of Priyadarśī's domains.

Of course, the proposal that Aṁtiyoka was a minor Indian potentate is tantamount to a major paradigm shift in modern historical thought. Before we can seriously contemplate such a shift, we will have to consider much additional evidence. The next step is to examine how the five names were discovered and try to evaluate how close they are to the names of the five celebrated Hellenistic kings.

II. AṀTIYOKA

Aṁtiyoka is the best attested of the five names. The following table from Schneider (1978, pp. 25, 76) lists the forms of this name that were known from the rock edicts as of 1978.

Er	aṁtiyoke	aṁtiyokenā	[aṁtiyo]ge	aṁtiyogasa
Ka	atiyoge	a[ṁ]tiyogenā	aṁtiyoge	[a]ṁtiyogasā
Ma	...tiyo[ge]		[a]tiyoge	...[gasa]
Sh	aṁtiyoko	atiyok[e]na	aṁtiyo[k]o	Aṁtiyokasa
Gi			aṁtiyako	aṁtiy[a]kas[a]
Dh			...[t]iyoke	aṁtiyo[ka]sa
Jg			aṁtiyoke	Aṁtiyokasa

Table 4. Forms of Aṁtiyoka found in rock edicts 2 and 13. Letters that were difficult to read are enclosed in square brackets. Two-letter codes designate the rock edict sites where the names are found. Er: Erṛdaguḍi; Ka: Kālsī; Ma: Mānsehrā; Sh: Shāhbāzgarhī; Gi: Girnār; Dh: Dhauli; Jg: Jaugaḍa.

The letters enclosed in square brackets proved difficult to read. The Aśokan rock edicts were generally written in a careless fashion, and the effects of time have reduced their original legibility. Figure 1 illustrates how the five names look in one of the original inscriptions. The figure shows the line from the Kālsī rock edict 13 giving the names of the five kings. This edict is written in the Brāhmī alphabet, which is based on 35 letters and a system of vowel marks similar to the one used in Devanagari. The reader might like to try reading this line using the Brāhmī alphabet table in Appendix 3. (The answer is given in Appendix 4.)

[Scanned line from Kālsī rock edict 13]

Figure 1. Line from the Kālsī rock edict 13 giving the names of the five kings. This is taken from a rubbing of the inscription published by Bühler (1894).

From the information in Table 4, Schneider concluded that the proper form of the name Aṁtiyoka is Aṁtiyoke or Aṁtiyoge. This name is certainly similar to the Greek name Antiochos, but can we conclude that it actually refers to one of the Seleucid kings of that name? Before reaching a final conclusion, we should first examine the other four names.

III. TURAMAYA

In his narrative about the Shāhbāzgarhī inscription, Masson (1846, p. 296) made a passing reference to the villages of Mirdān, Hotti, Tūrū, and Meyār near

Shāhbāzgarhī. Note the similarity between Tūrū-Meyār and Turamaya. "It's just a coincidence," you might say. Perhaps so, and I wonder how often such coincidences show up in the vast collection of Asian, African, and European names. The question is: Could the alleged correspondence between Turamaya and Ptolemaios also be coincidental, or is Aśoka's Turamaya identical with one of the Ptolemies of Egypt?

To Prinsep, identity was proven by the juxtaposition between the name Turamaya and the word Yona-rāja, which he associated with king Antiochus the Great. Prinsep noticed that Turamaya followed shortly after this word. He said,

> "The sight of my former friend the *yona rāja*, (whom, if he should not turn out to be Antiochus the ally, I shall soon find another name for), drew my particular attention to what followed; and it was impossible, with his help, not to recognize the name of Ptolemy even in the disguise of Turamayo" (Prinsep, 1838, p. 225).

His implicit reasoning was that if Turamaya is mentioned in connection with a Hellenistic king, then instead of seeking a match for this name in the total set of ancient names, we should restrict our attention to names of Hellenistic kings. Once this is done, it is obvious that Turamaya must be one of the Ptolemies of Egypt.

This reasoning can be accepted, as long as we realize that the key element here is the identification of Aṁtiyoke the Yona-rāja with one of the Hellenistic kings named Antiochus. Without this identification, there is no compelling reason to link Turamaya with Ptolemaios, since the resemblance between these two names is not very great.

Prinsep tried to strengthen his case by suggesting that the 'r' in Turamaya was doubtful, and the actual reading of the inscription might be Tulamaya, which is much closer to Ptolemaios. It turns out that scholars have continued to read this letter as an 'r' in the Girnār inscription. However, in the Kālsī and Erṛdaguḍi rock edicts the word is written Tulamaye.

This might seem to vindicate Prinsep, and Schneider (1978, p. 76) seems to agree. But it turns out that the Prakrit dialects of the Kālsī and Erṛdaguḍi edicts convert *all* r's into l's. For example, *rāja* becomes *lāja*. One can argue that the original name Turamaya was converted into Tulamaya when it was written in these dialects. The dialects using the spelling Turamaya (at Girnār and Shāhbāzgarhī) used both r's and l's. So if the name was originally Tulamaya, there would be no reason to write it as Turamaya in these dialects.

The elements of Turamaya appear in Sanskrit literature. In the *Bhāgavata Purāṇa* (9.22.38) there is a reference to Tura, the son of Kalaṣa, a priest of

Janamejaya. *Tura* means speedy or energetic in Sanskrit. There are also many Purāṇic references to an Asura named Maya who is famous for his technological expertise. Thus Turamaya could possibly be a native Indian word.

Curiously enough, the resemblance between Turamaya and Asura Maya has been exploited by indologists. Ebenezer Burgess, in his translation of the *Sūrya-siddhānta,* maintained that the Asura Maya of that text is a corruption of Turamaya – here identified as the astronomer Ptolemy of the second century A.D. (Burgess, 1860, p. 4).

IV. AṀTEKINI

Prinsep first read the name Aṁtekini in the Girnār rock edict 13 as Gongakena, but he conjectured that the correct reading should be Antikono, representing the Macedonian king Antigonus Gonatas (Prinsep, 1838, pp. 224–25). Later scholars only partially confirmed this guess, giving the following readings (Schneider, 1978, p. 76).

Site	Reading
Erṛdaguḍi	aṁt[i]k[e]ni
Kālsī	aṁteki[ne]
Mānsehrā	aṁt[e...
Shāhbāzgarhī	aṁtikini
Girnār	[a]ṁt[ek]ina

Table 5. Forms of Aṁtekini.

As before, the brackets around some of the letters indicate that they are difficult to read. Aṁtekini seems to be poorly written in several of the inscriptions, and its exact spelling is therefore somewhat uncertain.

(a).xxxxx (b).xxxxx (c).xxxxx (d).xxxxx

Figure 2. Successive facsimiles of Aṁtekini from the Girnār inscription. These are from (a) Prinsep (1838), (b) Jacob and Westergaard (1843), (c) Wilson (1850), and (d) Bühler (1894).

Prinsep originally read this name in a facsimile of the Girnār inscription in which the first letter looked like a poorly written Brāhmī 'go' (Figure 2a). Since the third letter looked like a 'ga', he arrived at Gongakena. 'Go' can be seen as a Brāhmī 'a' on its side, and 'ga' could be the lower half of 'ta' with some unknown vowel marking. This suggests a reading such as Aṁtakena, and Prinsep therefore proposed Antikono in hopes of matching the Greek name Antigonus.

In two later facsimiles reported by Jacob and Westergaard (1843) and Wilson (1850), the third letter of the name is still written as a 'ga', as it was in Prinsep's facsimile (Figures 2b, 2c). Jacob and Westergaard specifically commented that it is "very doubtful if there be an upper stroke to constitute a ta." However, in the estampage published by Bühler (1894), the upper stroke and vowel mark of a 'te' are clearly visible (Figure 2d). The initial 'a' is very poorly formed, and one can see how it might once be mistaken for a 'go'. But it is surprising that three successive observers would read a clear 'te' as a 'ga' – unless, of course, the inscription was deliberately modified to improve the evidence.

But let us leave aside the question of data improvement. The accepted readings of Aṁtekini in Table 5 are not very similar to the Greek name Antigonus or to Prinsep's Antikono. Yet in Prakrit it is easy to spell and pronounce Aṁtikono and Aṁtigono. Bühler argued that Aṁtekini matches the Greek name Antigenes, but unfortunately none of the Hellenistic kings had this name (Bhandarkar, 1955, p. 43). As with Turamaya, the identification of Aṁtekini with a known Hellenistic king seems to depend on the precedent set by the initial identification of Aṁtiyoke as Antiochus.

V. MAGĀ

The rock edicts clearly record the name Magā or Makā. If Turamaya and Aṁtiyoke are accepted as Hellenistic kings, it is natural to seek such a king with a name resembling Magā. Prinsep (1838, p. 225) pointed out that Ptolemaios Theos had a half-brother named Magas who governed Cyrene, a small Egyptian territory in what is now Libya. The name Magas is certainly similar to Magā, but Magas was a very obscure and unimportant ruler. Why would he be singled out for mention by Aśoka among so many other potentates of Asia, Africa, and Europe?

There are many names in India making use of the word *maga*. For example, there are the Magā brahmins and the Magadhas, or bards, after whom the province of Magadha was named. There is even a Makām river near Shāhbāzgarhī (Woolner, 1924, p. xi). Were it not for the Western identifications of Aṁtiyoke, Aṁtekini, and Turamaya, one would not be justified in going all to way to Libya to look for Magā.

By the way, there was a Carthaginian general named Mago who lived in about 550–500 B.C. (Davis, 1861, p. 96). Another Carthaginian general named Mago died in the second Punic war in 203 B.C. (Kinder, 1974, p. 83). If Mago is an ancient Phoenician name, then it might show up in many widely separated times and places.

VI. ALIKASUDARE

Although Prinsep was the first to read and identify four of the five Greek names, he missed Alikasudare since it was not present in the Girnār or Dhauli inscriptions. This name was first revealed in the Shāhbāzgarhī inscription, which was discovered by M. Court in 1836 (Masson, 1846, p. 293). In 1838, C. Masson visited the site of the inscription in what is now northern Pakistan and made copies and inked impressions on calico cloth.

The inscription is written in the Bactro-Pali or Kharoṣṭhī alphabet and was initially difficult to record and read properly. It was first deciphered by Norris and Dowson in 1845. Norris commented repeatedly in his report to the Royal Asiatic Society that the inscription was poorly legible (Norris, 1846). This was especially true of the back side of the rock where edict 13 was inscribed. However, he noted that "from this illegibility one line, containing the names of the five Western Kings must be fortunately excepted" (Wilson, 1850, p. 156). The unusual legibility of this line gives rise to some questions which I discuss in Appendix 1. The line itself was read as follows in 1850:

"Antiyoko nama yona raja parancha tena Antiyokena chaturo | | | |
rajano Turamara nama Antikona nama Mako nama
Alikasunari nama" (Wilson, 1850, p. 225).

Later on, students of the Shāhbāzgarhī inscription made some revisions in the spelling of these names. They decided that Turamara should be Turamaye, Antikona should be Aṁtikini, Mako should be Maka, and Alikasunari should be Alikasudaro. Note that Antikona is much closer to Antigonus (and Prinsep's Antikono) than Aṁtikini.

The name Alikasunari was inevitably interpreted as Alexander, and Norris was perhaps the first to suggest that this might be King Alexander of Epirus, a tiny principality near what is now southern Albania (Norris, 1846, p. 305). Since then, scholars have generally identified Alikasudaro either with this Alexander or with Alexander of Corinth, a city in the Peloponnesus of Greece.

In books on Aśoka, Alikasudaro has sometimes been written as Alikasundara or Alikasuṁdara, even though this spelling has apparently not been observed in the inscriptions (Ojha, 1968, p. 59, and Bhandarkar, 1955, pp. 43, 45). This change – an example of data improvement – does make Alikasudaro sound more like Alexander, but the resemblance is still not very close. The Sanskrit form of Alexandros should be something like Alakṣandraḥ. In Prakrit, this should become Alaṣanda or Alakkhanda. In comparison with these forms, Alikasuṁdara and Alikasudaro both stand out as anomalies.

The word Alasandā is found in Buddhist literature as a name for a city in the land of the Yonas. The Buddhist chronicle *Mahāvaṁsa* says that there was a large Buddhist community there. It goes on to say that once "the thera Yonaka Mahā Dhammarakkita came to Anurādhapura [in Sri Lanka] from Alasandā with 30,000 monks" (Malalasekera, 1960, p. 187). This Alasandā is accepted by the Russian scholar Bongard-Levin as the Alexandria founded by Alexander the Great near Kabul (Bongard-Levin, 1985, p. 242).

In the Pali text *Milinda* there is a reference to an Alasanda which Tarn (1951, p. 420) takes to be Alexandria of the Caucasus. Thus the form Alasanda is attested at least twice as an actual Prākrit word for Alexandria. This suggests that Prakrit speakers would not represent Alexandros as Alikasudaro.

Interestingly enough, the word Alikasuṁdara is meaningful in Sanskrit. *Alika* means false, and *suṁdara* means beauty. Thus Alikasuṁdara could mean "deceptive beauty." Since *alaka* means hair, the word Alakasuṁdara could also be a name meaning "one with beautiful hair." It is therefore plausible that Alikasudaro may derive from a Sanskrit name and may have nothing to do with Alexander.

Note that Prinsep's method of proof by association works in reverse. According to this method, if Aṁtiyoke is Antiochus, then the other names listed along with Aṁtiyoke must also refer to Hellenistic kings. But likewise, if Alikasudaro refers not to Alexander but to someone living in India, then one can argue that the other names listed with it also refer to people living in India. We have seen that Turamaya, Aṁtekini, and Alikasudaro are not very close to Ptolemaios, Antigonus, and Alexander. Magā is close to the name of the obscure ruler Magas, and Aṁtiyoke is close to Antiochus. The question is, do the strong points of Aṁtiyoke and Magā counteract the weak points of the other three names, or is it the other way around?

It should also be pointed out that the name Alexander was quite old in the days of Alexander the Great. Paris of Troy was also named Alexander, and some scholars believe that he corresponds to king Alakshandush of Vilusha in Asia Minor, who lived in about 1300 B.C. (Nilsson, 1968, p. 105). If Alikasudaro matches Alexander, then it also matches Alakshandush and other similar royal names scattered over the centuries. It is hard to say when and under what circumstances these names might have found their way to India.

VII. SAHADEVA VISITS ANTIOCH AND ROME

Thus far, we have seen that (1) historical links between Aśoka and the Hellenistic world are lacking and (2) some of the accepted identifications of the

"Greek" names in Priyadarśī's edicts have serious shortcomings. It therefore makes sense to seek an alternative hypothesis to account for these names. To prepare for this, I will first take a side-excursion to South India with Sahadeva of the famous Pāṇḍava brothers.

In the *Mahābhārata*, it is said that when king Yudhiṣṭhira of Hastināpura wanted to perform the Rājasūya sacrifice, he sent his four brothers to conquer the surrounding kings in the four directions, north, east, south, and west. Sahadeva was sent to the south, and after campaigning for some time and subduing many kingdoms, he reached the region of the Pāṇḍyas and Tamils in South India. In the translation of van Buitenen (1975, p. 84), this is described as follows:

> "Likewise by means of envoys he subjugated and made tributary the Pāṇḍyas and Tamils, Coḍras and Keralas, Āndhras and Talavanas, Kalingas and Uṣṭrakarṇikas, Antioch and Rome, and the city of the Greeks."

Antioch and Rome? How did Sahadeva wind up in these cities, which lie far to the west of Hastināpura? Did the author of the *Mahābhārata* really think these cities are in South India? Or did he think Sahadeva made a sudden side-trip to the Mediterranean while on his southern tour?

The mystery deepens when we look at the Sanskrit for "Antioch and Rome, and the city of the Greeks." In the critical edition of the *Mahābhārata* (*Sabha Parva*, chap. 28, verse 49), this reads,

"antākhīṁ caiva romāṁ ca yavanānām puraṁ tathā"

The word translated as Antioch is Antākhīṁ. Now it turns out that Antakiya is the modern name of Antioch (Downey, 1961, p. 3). Antioch was founded in 300 B.C. by Seleucus Nikator, and it flourished as an important city until the Muslim conquest of Syria in the 7th century. Its fame was somewhat revived during the Crusades and in late Byzantine times, but it was reduced to obscurity under the Turks. In modern times the small town on the site of Antioch's ruins is called Antakiya.

So how did that name get into the *Mahābhārata*? Some insight into this matter is given by B. S. Suryavanshi (1986, p. 21). He points out that while Antākhīṁ appears in the critical edition of the *Mahābhārata*, the words Āvarim, Ashṭavīm, and Āṭivīm are used in place of Antākhīṁ in a number of other texts. He argues that Āṭivīm refers to the Āṭavikas who lived near Kaliṅga or modern Orissa. This is consistent with Sahadeva's itinerary, in which he turned north from Tamil Nadu, conquered the Āndhras, Talavanas, Kaliṅgas, and Uṣṭrakarṇikas, and then reached "Antioch and Rome" (Suryavanshi, 1986, p. 29). This would put "Antioch and Rome" somewhere near the lands of the Āṭavikas in the vicinity of Kaliṅga.

According to Bhandarkar (1955, p. 42), the Āṭavyas or Āṭavīs are mentioned in the *Purāṇas* along with Pulindas, Vindhyamūlīyas, and Vaidarbhas. There is a copper-plate grant describing a king Hastin, master of the Dabhālā kingdom together with 18 forest kingdoms or Āṭavī-rājya. Bhandarkar says that Dabhālā must be modern Bundelkhaṇḍ. He suggests that in the Gupta period the Āṭavī country must have extended from Bāghelkhaṇḍ almost to the sea-coast of Orissa.

Suryavanshi maintains that it is hard to see why the editor of the *Mahābhārata*, Franklin Edgerton, selected Antākhīṁ for the critical edition, rather than Āṭivīm. Of course, one possibility is that Edgerton perpetrated a hoax. But assuming that this is not the case, why did the word Antākhīṁ appear in even one manuscript of the *Mahābhārata*? Did an earlier redactor of the text interpolate a name for Antioch out of ignorance or some bizarre motive? Or could it be that Antākhīṁ is another obscure name of the Āṭavikas or of some people living near the Āṭavikas?

What about Rome (Romāṁ) and the city of the Greeks (Yavanānām Puraṁ)? One could argue that three references to the Mediterranean world reinforce one another and cannot be denied. However, some of the manuscripts of the *Mahābhārata* refer to Ramyaṁ instead of Romāṁ (Suryavanshi, 1986, pp. 28–29). In addition, there are several Sanskrit names beginning with *roma* (hair), such as Romaharṣaṇa (hair standing on end), Romapāda, and Romaśa. The *Viṣṇu Purāṇa* mentions Romāṇas in a long list of names of peoples (Wilson, 1989, p. 278). The Gypsies of Europe originated in India and call themselves Romany. It seems doubtful that these names all derive from contact with the Romans. Quite possibly they and Romāṁ are simply names of Indian origin that coincidentally resemble Roma. The terms Romāṁ and Ramyām may refer to people living in the vicinity of the Āṭivīm.

Regarding Yavana Puri, many scholars, such as H. H. Wilson (1989, p. 280), insist that the word Yavana originally referred to the Greeks. However, many Sanskrit texts use this word to refer to a class of people who were originally Aryan, but who had deviated from Aryan culture. For example, Wilson's translation of the *Viṣṇu Purāṇa* describes how the sage Vasiṣṭha separated the Yavanas "from affinity to the regenerate tribes, and from the duties of their castes" in order to save them from the wrath of king Sagara (Wilson, 1989, p. 536). Here the phrase "regenerate tribes" designates the followers of *varṇāśrama*, a social system that later developed into the modern Indian caste system.

Clearly the Greeks, as we know them historically, never followed the Indian *varṇāśrama* system. If the word Yavana originally referred to the Greeks, then one would have to suppose that the Purāṇic traditions tracing the Yavanas to the *varṇāśrama* system were invented later on. One can always hypothesize this, but it is also possible that these traditions are genuine.

There are also Purāṇic references to the existence of Yavana kingdoms that are *not* west or north-west of India. For example, the *Bhāgavata Purāṇa* commentator Viśvanātha Cakravartī relates a story from the *Viṣṇu Purāṇa* about Kālayavana, the son of a Yavana king living to the south of the Yādavas (Bhaktivedanta Swami, 1988, p. 227). This Kālayavana was "as black as a bee," which is not typical of the Greeks or Romans (Wilson, 1989, p. 783). Since the Yādavas were based in Mathurā, this sets a precedent for the presence of Yavanas somewhere to the south of that city. Thus it is plausible that Sahadeva could have encountered a Yavana city in his southern campaign. (Curiously, H.H. Wilson (1989, p. 783) sets the story of Kālayavana "on the shores of the Western sea," even though the word *dakṣina* or southern is used in the text.)

VIII. AN ALTERNATIVE IDENTIFICATION OF THE FIVE KINGS

King Priyadarśī begins his 13th rock edict by regretting the suffering and loss of life caused by his invasion of Kaliṅga. Then he mentions the forest peoples or Aṭavi (Woolner, 1924, p. 55). The Aṭavi, he says, have also accepted his ideals, but he warns them that he retains the power to punish wrongdoers, despite his remorse over his subjugation of Kaliiga (Nikam, 1966, pp. 28–29).

Aśoka then makes his famous statement about the five kings. Although I have quoted this already, I reproduce it here for ease of reference:

> "King Priyadarśī considers moral conquest [*Dharma-vijaya*] the most important conquest. He has achieved this moral conquest repeatedly both here and among the peoples living beyond the borders of his kingdom, even as far away as 600 *yojanas*, where the Yona king Antiyoka rules, and even beyond Antiyoka in the realms of the four kings named Turamaya, Antikini, Maka, and Alikasudara, and to the south among the Cholas and Pāṇḍyas as far as Ceylon" (Nikam, 1966, p. 29).

An interesting parallel can be seen between these passages in Aśoka's edict and the story in the *Mahābhārata* of Sahadeva's southern campaign. The left hand column of Table 6 lists the places visited in the last leg of this campaign, when Sahadeva was turning north from Tamil Nadu. Priyadarśī seems to be mentioning the same peoples listed in Sahadeva's itinerary, but he is going from north to south rather than from south to north. To show this, I have listed the peoples and kings mentioned by Priyadarśī in reverse order in the right hand column of the table.

Both columns of the table begin by mentioning kingdoms in the southern tip of India. Going north, they both reach Kaliṅga (modern Orissa). In the vicinity of Kaliṅga they both refer to the Āṭavī. In this same vicinity, they also refer to what scholars take to be Rome, a Greek city, and Antioch (*Mahābhārata*), and a famous king of Antioch (Aśokan rock edict 13). The last four lines of the table do not perfectly line up, but they all seem to refer to the same geographical region (in and around Orissa).

In the case of Sahadeva's southern tour, the references to Greeks, Romans, and Antioch simply make no sense if we take them literally. However, we have seen that the Sanskrit words in question may really refer to Yavanas who could be associated with the kingdoms of the Āṭavīs.

Sahadeva's campaign	**Rock edict 13 in reverse**
	Ceylon
Pāṇḍyas and Tamils	Pāṇḍyas
Coḷras and Keralas	Cholas (Coḷa)
Āndhras and Talavanas	
Kaliṅgas and Uṣṭrakarṇikas	four neighboring kings
Antākhī (Antioch?)	Aṁtiyoke (king Antiochus Theos?)
Āṭavī (in some mss.)	Āṭavī
Rome and Greek city?	Kaliṅga

Table 6. Peoples and kings mentioned in Sahadeva's southern tour and in rock edict 13. The order in rock edict 13 is reversed for the sake of comparison with Sahadeva's tour.

Could it be that the same is true of Aśoka's rock edict 13? Table 6 suggests that Priyadarśī is listing peoples and kings from north to south. But if Aṁtiyoke and company are Hellenistic kings, then the names in edict 13 jump mysteriously from Kaliṅga and the Āṭavīs, to the Mediterranean region (as far as Libya), and from there to South India. Since Priyadarśī is simply giving a list, this is not as bad as saying that Sahadeva visited Antioch and Rome on a march from South India to Orissa. But the parallelism between the two columns of the table suggest that Priyadarśī's "Greek kings" correspond to Sahadeva's "Greeks and

Romans." If the latter are native Indian Yavanas living near Orissa, then the same should be true of the former.

But why should Yavanas living near Orissa have names that remind us of Greeks and Romans? I would suggest that we are dealing here with coincidental similarities between words. The following table lists the key words that we have been considering:

1.	Alikasudaro	Alexandros
2.	Aṁtekini	Antigonus
3.	Turamaya	Ptolemaios
4.	Magā	Magas
5.	Aṁtiyoke	Antiochus
6.	Antākhī	Antakiya (in Turkey)
7.	Romāṁ	Roma (in Italy)

Table 7. Pairs of apparently related words.

It may appear that each word in the left column is derived historically from the corresponding word in the right column. But I have given several reasons for doubting this: First of all, Alikasudaro in line 1 is probably not the Prākrit form of Alexandros. If Alikasudaro is not a Hellenistic king, then the matches in lines 2–5 are cast into doubt. The matches in lines 2 and 3 are not very close to begin with. Magā in line 4 is close to Magas, but it seems odd that Aśoka would single out Magas of Cyrene in Libya for special mention. The same can be said of Alexander of Epirus or Corinth.

Someone might argue that Antākhī in line 6 is a recent name for Antioch that somebody inserted into the *Mahābhārata,* but this theory has some drawbacks. If this insertion occurred before Prinsep's reading of Aṁtiyoke in 1838, then it involves some remarkable coincidences. First of all, the interpolator selected Antioch out of the total set of ancient cities without knowing that Prinsep would later find an Antiochus in Aśoka's inscriptions. In addition, he placed this Antioch in Sahadeva's tour so that it would line up geographically with the later-to-be-discovered Antiochus as shown in Table 6. Of course, the insertion may have been made after 1838 by someone who knew of Prinsep's discoveries. But why would such a person insert a *modern* form of Antioch? And why not put Antioch in Nakula's tour, which went to the west?

Variant texts of the *Mahābhārata* indicate that Antākhī is associated with the Āṭavīs of Orissa. This and the drawbacks of the recent insertion theory suggest that Antākhī has nothing to do with Antakiya (Antioch) by the Mediterranean sea.

There are many Sanskrit words using the syllables *roma*, and therefore the Romāṁ of line 7 probably has nothing to do with Roma in Italy. If the Antākhī and Romāṁ of lines 6–7 really do refer to the Mediterranean region, then Sahadeva's itinerary is rendered absurd. Thus if the *Mahābhārata* makes sense, the matches in lines 6–7 should be coincidental.

Here one could argue that Greek and Roman merchants used to have trading centers in South India. Perhaps Antākhī and Romāṁ refer to these people. But if this is true, then the parallelism shown in Table 6 suggests that Aṁtiyoke and his colleagues must also be in South India. This means that they are not Hellenistic kings ruling in the Mediterranean region.

Of course, one could say that the parallelism in Table 6 is coincidental. The Greeks mentioned by Priyadarśī have nothing to do with those mentioned in the story of Sahadeva. But one could just as well say that the match between Aṁtiyoke and Antiochus is coincidental. It becomes a game of "name your coincidence."

The two lines 5 and 6 might involve a sound change (dropping the 'yo' sound) that worked both in India and the West. It seems reasonable to suppose that the modern name Antakiya derived from the ancient Antiokheia through such a change, which took place over many years as pronunciations shifted. If Antākhī was not a recent import of Antakiya into the *Mahābhārata*, then it may have derived from Aṁtiyoke by a similar sound change that occurred over many years in India. If this is true, then Aṁtiyoke must have been used for many years in native Indian speech, and it probably does not refer to a foreign monarch such as Antiochus Theos, who would hardly have been widely discussed in India. Of course, another possibilty is that Antākhī is only coincidentally similar to Aṁtiyoke.

In either case, the name Aṁtiyoke in line 5 may be only coincidentally similar to Antiochus. Rather than being an emperor of Syria and Persia, Aṁtiyoke may have been a minor Yavana king (Yona-rāja) having something to do with Antākhī and the Āṭavīs. His four colleagues would have been similar minor kings ruling in the same general area.

It is instructive to compare the word pairs in Table 7 with the Sanskrit etymologies created by P. N. Oak (1973) for European place names. For example, Oak derives Thames from *tāmasa* (dark), since the Thames river in England is murky and runs through foggy country. He breaks down Scandinavia into Skanda (the god of war) plus *navi* (boat) – a reference to the warlike Vikings and their ships. Since it is easy to create such etymologies, it seems that coincidental similarities between European and Sanskrit names are not uncommon.

One objection to our alternative hypothesis is that after listing the four kings,

Aśoka's rock edict 13 says "and to the south among the Cholas and Pāṇḍyas." Some scholars have interpreted this to mean that Priyadarśī was indicating a change in direction from west to south. However, a reference to the south is not out of place if Priyadarśī was progressing from north to south in his list of places and names. In addition, the word *nica*, which is translated here as south, may have a different meaning. In Apte's Sanskrit Dictionary, *nīca* is said to mean low or vile, and no meaning is listed referring to the south. Thus Priyadarśī might have used *nica* to indicate that the peoples he was listing were lowly or degraded. We do find in an Aśokan Prākrit glossary that *nica* means "in the south," but that interpretation may be based on the accepted reading of rock edict 13 (Woolner, 1924, p. 103).

Here is another objection. Rock edict 13 says that Aṁtiyoke is ruling 600 *yojanas* beyond the borders of Priyadarśī's kingdom. At 4.5 miles per *yojana*, this comes to 2,700 miles, or roughly the distance to Antioch. Surely this supports the identification of Aṁtiyoke with Antiochus.

Curiously enough, one can read the edict in such a way as to give even stronger support to the standard view. The phrase, "beyond the borders of his kingdom, even as far away as six hundred *yojanas*," is ambiguous. It could mean "600 *yojanas* beyond the borders" or "600 *yojanas* away from here, out beyond the borders." The "here" in the latter reading would be Aśoka's capital of Pataliputra (modern Patna).

According to the *Sūrya-siddhānta* the earth's diameter is 1,600 *yojanas*, and the radius is therefore 800 *yojanas* (Burgess, 1989, p. 43). From the latitudes and longitudes of Patna and Antioch, we can compute the great circle distance between them in radians. If we multiply this by 800, we find that this distance is 598.9 *yojanas*. This is only 1.1 *yojanas* off from the 600 *yojana* figure in Aśoka's edict.

This looks like a remarkably good agreement. But if it represents real knowledge of the great circle distance between Pataliputra and Antioch, then both Greeks and Indians in the 3rd century B.C. must have made accurate measurements of latitudes and longitudes. This is contrary to the accepted history of geography. For example, it appears that the astronomer Ptolemy in the 2nd century A.D. did not have accurate knowledge of longitudes in India or even in the Mediterranean region (Nordenskiold, 1897). The agreement between 586.9 and 600 *yojanas* thus seems to be another one of those coincidences that keep popping up in this study.

The length of the *yojana* is highly variable, ranging from about 4.5 miles to 8 or 9 miles (Cunningham, 1990, pp. 483–89). The *Sūrya-siddhānta yojana* falls at the low end of this scale at 5 miles/*yojana*. One could argue the figure of 600 refers to a winding, circuitous route from Pataliputra to Antioch that was measured in larger *yojanas*. Or one could argue that this figure represents a winding route in India measured in smaller *yojanas*.

The 600 *yojanas* might refer only to the distance to Aṁtiyoke's capital, or it might refer to additional peoples and places "as far as Ceylon [Sri Lanka]." If we measure the overland distance from Patna to the southern tip of Sri Lanka along a reasonably direct route going down the east coast of India and then crossing Adam's bridge to Sri Lanka, we get a figure of about 1,700 miles. This comes to about 378 *yojanas* of 4.5 miles. For a winding path to total 600 *yojanas*, the path would have to be about 1.6 times as long. This is plausible if the path followed winding roads and went inland to visit the territories of various peoples. Thus the 600-*yojana* distance does not pose an insuperable obstacle to our alternative hypothesis.

A final objection might be that the standard reconstruction of Indian chronology makes Aśoka a contemporary of the Hellenistic kings of Table 1. Since their names and dates are well known, there is no justification in seeking alternative identifications of Aśoka's five kings among undated and poorly known peoples somewhere in India. The standard chronology is based on extensive scholarly studies, and it should be accepted as objective knowledge.

In reply, I should point out that it is beyond the scope of this paper to reassess the modern system of Indian chronology. However, it is true that the identification of Aśoka's five kings with Hellenistic monarchs is one of the important foundation stones of this system. To argue that the identifications should be accepted because the modern system is true smacks of circular reasoning. Such an argument can be accepted only if the modern system of chronology can be established without reference to the identification of the kings.

There are good reasons for thinking that the accepted identifications of Table 1 are incorrect – or at least highly questionable. At best, they are little more than conjectures based on intriguing similarities between words. On the negative side, these conjectures are burdened by the following drawbacks:

> (1) If Priyadarśī-Aśoka was actually in close contact with famous Hellenistic kings, there should be clear evidence of this in Western historical records. But no such evidence is known. Furthermore, the Buddhist chronicles describe Aśoka's efforts to spread Buddhism in countries near India, but they make no mention of missions to the Mediterranean region (section I).

(2) Turamaya, Aṁtekini, and Alikasudaro are not very close phonetically to Ptolemaios, Antigonus, and Alexander (sections III, IV, and VI). In particular, Alikasudaro is not the Prākrit word for Alexander.

(3) The accepted identifications of the five kings were fixed at a time when the Aśokan inscriptions were first deciphered and could not be clearly read. There is evidence that the Hellenistic leanings of the early decipherers influenced the initial reading of the names (sections IV and VI). There is even some evidence of questionable dealings in the decipherment of the Shāhbāzgarhī inscriptions (Appendix 1).

(4) The parallels between the story of Sahadeva's southern campaign in the *Mahābhārata* and Priyadarśī's rock edict 13 strongly indicate that Aṁtiyoka and his four colleagues were situated near Orissa rather than in the Mediterranean region (section VIII).

If the modern reconstruction of Indian history can stand without the accepted identifications of Aśoka's five kings, then historians would be better off without them. But if the modern reconstruction cannot stand without these identifications, then it rests on very insecure foundations and is in need of serious scrutiny.

APPENDIX 1. THE STORY OF THE SHĀHBĀZGARHĪ INSCRIPTION

The Frenchman M. Court discovered the Shāhbāzgarhī inscription in 1836 and mentioned it in a single line of a memoir on ancient ruins near Peshawar. Court reproduced 23 letters from the inscription, which he said were all he could read since the inscription was almost effaced by time (Court, 1836, p. 481). Later on, an agent of Court made a transcript which "appears to have embraced the whole rock" (Wilson, 1850, p. 156). Court sent this transcript to Professor Lassen in Germany, and he, in turn, sent a copy to the Royal Asiatic Society in England.

In October 1838, C. Masson arrived at the site of the inscription just after the departure of Court's agent (Masson, 1846, p. 298). He had learned of the inscription from one Captain Burnes, and he was led to believe that it consisted of only five lines (Masson, 1846, p. 293). After sending a native servant to make cloth impressions of the inscription, he learned that it was very extensive. Since the servant's work was of very poor quality, he mounted an expedition Shāhbāzgarhī to make a good facsimile.

Masson proceeded to clear the inscription of moss and make two cloth impressions of the letters inscribed on the northern or superior face of the rock. He said that the tilt of the rock and ground surface on the southern or back face made it impossible to make cloth impressions on that side. After clearing the letters with sharp metal tools and marking them with white, chalk-like stones, he made a careful copy of the inscription on the back side (Masson, 1846, pp. 299–300).

Masson read his report on the Shāhbāzgarhī inscription at a meeting of the Royal Asiatic Society in mid-January of 1845. This gave a detailed account of his adventures in the field, but it did not mention the possibility that he might be dealing with Aśokan edicts. On March 1, 1845, E. Norris read a report to the society describing how he and Mr. Dowson had deciphered the inscription. This work was based on the materials supplied by Masson.

Norris first made use of Masson's second cloth impression of the northern face, and later gained further insight from his first cloth impression. When he tried to decipher the inscription on the back of the rock he "was deprived of the resource of a cloth impression," but fortunately the line in Masson's transcription giving the names of the alleged western kings was "with one exception, perfectly legible" (Norris, 1846, p. 304).

In 1849, H. H. Wilson presented a paper on the Aśokan rock inscriptions to the Royal Asiatic Society. There he reproduced some further comments by Norris about the decipherment of the Shāhbāzgarhī inscription. In those comments, Norris amplified on his remark about the legibility of the line containing the names of the five kings. He said that, "from this illegibility one line, containing the names of the five Western Kings must be fortunately excepted, which Mr. Masson copied with special care, and even took off a cloth impression of a small portion, in spite of the difficulties presented by the position of the rock" (Wilson, 1850, p. 156).

This small cloth impression is shown in Norris's lithograph of the inscriptions at the beginning of Wilson's article. The impression is indeed small. It consists of the letters "*turo* | | | | *rajani tu*" from the middle of the line giving the names of the kings. Apart from some disconnected letters, this says "| | | | kings" or "4 kings."

Here we come to a part of the Shāhbāzgarhī story that strikes me as very puzzling. What motivated Masson to go to the trouble of making an impression of these particular letters? If he simply picked a spot at random on the back of the rock and transcribed a few of the letters there, it is highly unlikely that he would hit the center of the kings line. It would also be pointless to transcribe a few letters at random. It seems more reasonable to suppose that he could read the letters and that he made an impression of them because he thought they were significant. I would suggest that he knew they referred to the "4 kings" of Aśoka's rock edict 13. Indeed, since he copied the names of the kings with "special care," it would seem that he could read and understand all five names.

If Masson had openly stated this in his report to the Royal Asiatic Society, it would not be surprising. Masson was a regular participant in the affairs of the society, and he must have known about Prinsep's celebrated discovery of the names of Hellenistic kings in Aśokan inscriptions (Prinsep, 1938a and 1938b). It would have been reasonable for him to suppose that the Shāhbāzgarhī inscriptions might be Aśokan and to search through them for the names of the kings. When he spotted "| | | | *rajani*" he might well have thought "Here they are!" and made a cloth impression of these letters as evidence of his discovery. Then he might have deciphered the names and carefully copied them.

But Masson said nothing about this in his report. Indeed, he didn't even say that he thought he might be dealing with an Aśokan inscription. The question is: Why?

A month and a half after Masson's report to the Royal Asiatic Society, Norris and Dowson reported their successful decipherment of the Shāhbāzgarhī inscription. Norris explained that his knowledge of legends on Bactrian coins enabled him to recognize that the word *piyasa* occurred repeatedly in the inscription. This word was always preceded by three letters which he later identified as 'de' 'va' 'na' (Norris, 1846, p. 303).

This, according to Norris, was the first step in decoding the Shāhbāzgarhī inscriptions. But if Masson could already understand the names of the alleged Western kings, this cryptographic effort looks like a sham. At the very least, Masson or Norris should have reported that an important part of the inscription had already been deciphered in the field.

Could it be that Masson actually did not decipher the names of the kings? This is doubtful. The unusual legibility of the king's line indicates that he did decipher them. The line is,

> *"Antiyoko nama yona raja parancha tena Antiyokena chaturo | | | |*
> *rajano Turamara nama Antikona nama Mako nama Alikasunari nama."*
> (Wilson, 1850, p. 225)

From 1850 to 1924, only 8 changes were made in the reading of this line: 6 changes in vowel signs (as in Antikona to Antikini) and 2 changes in consonants (as in Turamara to Turamaye). This could be regarded as minor editing, especially since the vowel signs in the Kharoṣṭhī alphabet are small and easy to misread. In contrast, well over half of the letters in the remainder of rock edict

13 were changed between 1850 and 1924, and these were mostly changes in consonants. This indicates that the 1850 reading was largely erroneous, except for the line listing the five kings. For a detailed discussion of this, see Appendix 2.

The unique legibility of the kings line begins with the first 'A' of Antiyoko, and it ends with the last 'ma' of Alikasunari *nama*. Evidently Norris was correct when he said that Masson copied this line with "special care." I show in Appendix 2 that there is about 1 chance in 500,000 that this special care would begin at random with the first letter of this line and end with the last letter. Therefore, it seems reasonable to suppose that Masson understood the kings line and transcribed it carefully because he knew its importance.

Regarding the names of the kings, Wilson said, "It luckily happens that M. Court's copy is also very legible in this passage, and entirely confirms Mr. Masson's readings" (Wilson, 1850, p. 230). This might be taken to indicate that the kings line just happened to be more legible than the rest of the inscription on the back of the rock. Court's version of this line is lithographed in Wilson's article, and although it contains a number of garbled letters, one can still make out the names of the kings (excepting Aṁtiyoke). However, it is by no means as legible as the kings line that Norris obtained from Masson.

It seems clear that Masson deciphered the names of the five kings in the field but said nothing about this publicly. Norris appears to have known this, but he said nothing about it when he reported the decipherment of the inscription by himself and Dowson. The question is: Why was this secrecy necessary? Did Masson and Norris have something to hide? From the available evidence, it is impossible to know. However, this evidence does cast a shadow of doubt on the scholarship involved in the decipherment of the Aśokan edicts.

APPENDIX 2.
ON THE LEGIBILITY OF THE SHĀHBĀZGARHI INSCRIPTION

To evaluate the decipherment of the Shāhbāzgarhī rock edict 13, I compared two texts of this edict. The first, published by Woolner (1924), was chosen to represent mature scholarly conclusions about the edict. The second, published by Wilson (1850), represents the initial decipherment of the edict by Norris and Dowson in 1845.

The 1850 text consists of a series of blocks of letters separated by gaps representing unreadable sections of the inscription. With the aid of a computer, I lined up these blocks with the 1924 text in such a way as to give the best match. The differences between the 1924 text and the aligned 1850 text indicate changes which have been made in the decipherment between 1850 and 1924. Presumably, these are mostly improvements.

xxxxxxxxx**xxxxxa**xxxxxxxxxxxxaxxxxxxxxxxxxnxxxxxxxxxxxxxxxxxxxxxxxxxxnxxxxxxxxxxxxxx
xxxxxxxxxxxxxxxxa**xxxxxx**xxxxxxxxxx**xxxxxx**xxxxxxxxxxxxxxxxxxxxxxxxxxxxxxxxxxxxxx
xxxxxaaxnxaaxx**aaaaxaa**axaxxxxxxxaaax**xxxxxxa**xaxaaxxaxxxxxxxxnxxaxxxxxaaxxnaxaxxxxnxax
aaaaxaaxnxnxaxaxaaxxxnaxxxxxanaaaaxaaaaxxxaaaxxxxnxxxxanaaaxxaanaanxaaxxxaxxnaxxanxx
xxaaxxxaxxxxxxanxaxxaaxaaaxaxxaaaaxaaxxaaananxaxxxaxxaxaxxxxxnxxxxxxnxaxxa**aaaaaaaa**x
axxxaxxxxxaxxxaxxnxnaaxxnxanxnaxxxxaxxxxxxxaaanaaaaaaaaaxnxxxxx**aaaaxaa**nnxaxaxnaaxax
xaxnn**aaaaxaaa**xaxxaxxaaxxxx**aaaaxaa**xxnnxnaxnxaanxxnaaxxnaxxan**aaaaaaa**naanxaaxaxnnax
xxxanaaaa**aaaaana**xaaaxxaaaxaaxaaxxaxxaanxxxxxxxnaxnaaaaxaaaxaaaanaxxxx**aaaaxaa**xxxxxx
xxxnnxxxxaaaaaxxxaxx[**aaaaaaaaaaaaaaaaaaaaaaaanaaanaaaxaaaaannaaanaaaaaaxnaa**]
xxxxxxxaaxxxxxxxxxxxxaaxxxaxaaaxxnnxxxxxaxxaxaxnxaaxnxxnaaxx**aaaaaaa**xaaaaxaanxxxaxx
a**aaaaaaa**xxxxxaanaxnx**aaaaxaa**xanxaaaaxxaxxxxxxxxxaaaxaaxxanaxxxnxxxxnxaaaaaxaannxxxxa
anxaanxaaaxaaxaxxxxxxanxaxxaaxxxxxxaxxaaxx**aaaaxa**aanaxxxxxnxxxaxaaxxxxxaxnxaxaanxaan
anxaaxnxxxxaxxnxxxxxxxxxxxxnaxxnaaxxxxxaxaaaxxaxxxanaaxxxanxxxxaxaaxxxnxxxxxaaaaaaa

Figure 3. A listing of matches and mismatches between Norris and Dowson's 1845 decipherment of Shāhbāzgarhi rock edict 13 and Woolner's 1924 text of this edict. The letter 'a' designates an exact match, 'n' a near match, and 'x' a mismatch (see the text). The *devana-piyasa's* and the line giving the names of the five kings are marked in bold.

To graphically represent these changes, I selected three code symbols. The symbol 'a' represents an exact match between a letter in the 1924 text and the corresponding letter in the aligned 1850 text. The symbol 'n' represents a near match, in which the two letters have the same consonant value but differing vowel marks (e.g., 'ta' and 'te'). The symbol 'x' represents a mismatch. In this case, the two letters have differing consonant values (e.g.. 'kya' and 'ka') or differing values as pure vowels (e.g., 'a' and 'e'). The 'x' symbol is also used to represent a case where a letter in one text lines up with a gap in the other text.

The accompanying figure lists these symbols for the two aligned texts. To illustrate the process of decipherment, I put the symbols 'a', 'n', and 'x' in bold where the 1924 text has the words *devana piyasa* (in various spellings). As I pointed out in Appendix 1, Norris reported that the first step in decoding the Shāhbāzgarhī inscription was to recognize repeated appearances of these words (Norris, 1846, p. 303).

I also put the symbols 'a', 'n', and 'x' in bold in the 55-letter line representing the names of the alleged Greek kings, and I put square brackets around this line. (See section VI or Appendix 1 for this line.) This line begins with the first letter of Antiyoko, and it ends with the last letter of Alikasunari *nama*. Note that there are x's immediately before this line and immediately after it.

The text as a whole has 54.8% mismatches, 8.3% near matches, and 36.9% matches. Most of the matches and near matches are concentrated in small blocks, and several of these correspond to the name *devana-piyasa*. In contrast, the 55-letter line representing the names of the five kings has 2 mismatches, 6 near matches, and 47 matches.

Why was the line containing the names of the five kings so much more legible to early investigators than the rest of the text of edict 13? Did it happen by chance? There are over 1,000 letters in the 13th edict. We can compute the chances of randomly starting a line with the 'A' of Antiyoko and ending it with the 'ma' of Alikasunari *nama*. There are 499,500 ways of picking two letters out of 1,000, and so the chances of getting the starting and stopping letters exactly right are less than 1 in 499,500.

If this didn't happen by chance, perhaps it happened by some traceable cause. In an effort to understand what this cause might be, I carefully studied the history of the discovery, recording, and initial decipherment of the Shāhbāzgarhī inscription. My findings are presented in Appendix 1.

APPENDIX 3. THE BRĀHMĪ ALPHABET

[Place here a table of the Brāhmī alphabet, giving the 35 letters and the rules for adding vowel marks. A couple of examples of ligatures between consonants could also be given.]

APPENDIX 4. ANSWER TO THE EXERCISE

"Aṁtiyogenā catāli 4 lajāne Tūlamaye nāma Aṁtekine nama Makā nāma Alikyaṣudale" (Woolner, 1924, p. 29).

Bibliography

Bevan, E., *The House of Ptolemy*, Chicago: Argonaut, Inc., 1968.

Bhandarkar, D. R., *Asoka*, Calcutta: Univ. of Calcutta, 1955.

Bongard-Levin, G. M., *Mauryan India*, New Delhi: Sterling Publishers, 1985.

Bühler, G., "Aśoka's Rock Edicts According to the Girnar, Shāhbāzgarhī, Kālsī and Mansehra Versions," *Epigraphia Indica*, vol. 2, 1894, pp. 447–472.

Burgess, E., *The Sūrya Siddhānta*, ed. P. Gangooly, Delhi: Motilal Banarsidass: 1989 (first edition, 1860).

Cunningham, A., *The Ancient Geography of India*, Delhi: Low Price Publications, 1990.

Court, M. A., "Extracts translated from a Memoir on a Map of Peshāwar and the country comprised between the Indus and the Hydaspes, the Peucelaotis and Taxila of ancient geography," *Journal of the Asiatic Society of Bengal*, vol. 5, no. 56 (August), 1836.

Davis, N., *Carthage and Her Remains*, London: Richard Bentley, 1861.
Davis, N. and Kraay, C., *The Hellenistic Kingdoms*, London: Thames and Hudson, 1973.
Downey, Glanville, *A History of Antioch in Syria: From Seleucus to the Arab Conquest*, Princeton: Princeton Univ. Press, 1961.
Jacob, L. G. and Westergaard, N. L., "Copy of the Asoka Inscription at Girnar," *Jour. of the Bombay Branch of the Royal Asiatic Society*, 1, issue 5 (April), 1843.
Kinder, H. and Hilgemann, W., *The Anchor Atlas of World History*, Vol. 1, Garden City, New York: Anchor Books, 1974)
Majumdar, R. C., *The Age of Imperial Unity*, Bombay: Bharatiya Vidya Bhavan, 1951.
Majumdar, R. C., *Ancient India*, Delhi: Motilal Banarsidass, 1960a.
Majumdar, R. C., *The Classical Accounts of India*, Calcutta: Firma K. L. Mukhopadhyay, 1960b.
Malalasekera, G. P., *Dictionary of Proper Pāli Names*, Vol. 1, London: Luzac & Company, 1960.
Masson, C., "Narrative of an Excursion from Peshāwer to Shāh-Bāz Ghari," *Journal of the Royal Asiatic Society*, Vol. 8, Issue 15, January 1846.
Mukherjee, B. N., *Studies in Aramaic Edicts of Aśoka*, Calcutta: Indian Museum, 1984.
Nikam, N. A. and McKeon, Richard, eds. and trans., *The Edicts of Asoka*, Chicago: Univ. of Chicago Press, 1966.
Nilsson, Martin P., *Homer and Mycenae*, New York: Cooper Square Publishers, Inc., 1968.
Nordenskiold, A. E., *Periplus*, F. A. Bathev, trans., Stockholm: Norsted, 1897.
Norris, E., "On the Kapur-di-Giri Rock Inscription," *Journal of the Royal Asiatic Society of Great Britain and Ireland*, Vol. 8, pp. 303–307, 1846.
Oak, P. N., *Some Missing Chapters of World History*, Pune: P. N. Oak, 1973.
Ojha, K. C., *The History of Foreign Rule in Ancient India*, Allahabad: Gyan Prakashan, 1968.
Perry, Sir Erskine, "Account of the great Hindu Monarch, Asoka, chiefly from the Indische Alterthumskunde of Professor Christian Lassen," *Jour. of the Bombay Branch of the Royal Asiatic Society*, vol. 3, issue 14, Jan., 1851.
Prinsep, James, "Discovery of the name of Antiochus the Great in two of the edicts of Asoka, king of India," *Journal of the Asiatic Society of Bengal*, vol. 7, 1838a.
Prinsep, James, "On the Edicts of Piyadasi, or Asoka, the Buddhist monarch of India, preserved on the Girnar rock in the Gujarat peninsula, and on the Dhauli rock in Cuttack; with the discovery of Ptolemy's name therein," *Journal of the Asiatic Society of Bengal*, vol. 7, 1838b.
Prinsep, James, "Examination of the Inscriptions from Girnar in Gujarat, and Dhauli in Cuttack, continued," *Journal of the Asiatic Society of Bengal*, vol. 7, 1838c.
Smith, Vincent, *Asoka, the Buddhist Emperor of India*, New Delhi: S. Chand and co., 1964.
Schneider, Ulrich, *Die Grossen Felsen-Edikte Aśokas*, Wiesbaden: Otto Harrassowitz, 1978.
Sen, Amulyachandra, *Asoka's Edicts*, Calcutta: The Indian Publicity Society, 1956.
Suryavanshi, B. S., *Geography of the Mahabharata*, New Delhi: Ramanad Vidya Bhawan, 1986.
Tarn, W. W., *The Greeks in Bactria & India*, Cambridge: Cambridge Univ. Press, 1951.
Upham, Edward, *The Mahāvansi, the Rājā-Ratnācari and the Rājā-vali*, Vol. 1, London: Parbury, Allen, and Co., 1833.

van Buitenen, J. A. B., *The Mahābhārata, 2. The Book of the Assembly Hall, 3. The Book of the Forest*, Chicago: The University of Chicago Press, 1975.
Wilson, H. H., "On the Rock Inscriptions of Kapur di Giri, Dhauli, and Girnar," *Journal of the Royal Asiatic Society of Great Britain and Ireland*, vol. 12, 1850.
Wilson, H. H., trans., *The Viṣṇu Purāṇa*, Delhi: Nag Publishers, 1989.
Woolner, Alfred, *Asoka Text and Glossary*, Delhi: Rare Reprints, 1982 (original publication date, 1924).

CHAPTER 5

On the Antiquity of the Star Coordinates from Indian *Jyotiṣa Śāstra*

Essay presented at the
"International Symposium of Ancient Indian Chronology"
B. M. Birla Science Centre
Hyderabad, India / January 1994

I Introduction

A comparison is made between coordinates for 35 stars listed in traditional Indian astronomical texts (*jyotiṣa śāstra*) and the coordinates of corresponding stars listed in modern tables. I find that the error vectors pointing from the modern star positions to the corresponding *jyotiṣa* star positions are strongly correlated with the reversed proper motion vectors of the stars. Once precession is taken into account, the modern star positions show a tendency to move towards the *jyotiṣa* star positions as we go back in time.

To evaluate this, I first consider the null hypothesis, which says that we should not expect to find a significant relationship between errors in *jyotiṣa* star coordinates and proper motions of stars. I give statistical arguments showing that this hypothesis is not correct.

If there is a significant relationship between proper motions and *jyotiṣa* star coordinates, then the simplest explanation for this is that the *jyotiṣa* star coordinates were measured in the distant past. As time passed, the stars slowly moved from their positions and thereby generated error vectors pointing back along their paths. Given this hypothesis, it is possible to calculate the time of measurement of the *jyotiṣa* star coordinates. I find that these coordinates divide into a group 25,000–55,000 years old and a group less than 5,000 years old.

There is also a group that cannot be clearly dated, and there is evidence suggesting that the stars in this group may not be correctly identified.

II *Jyotiṣa* Star Coordinates

Several *jyotiṣa śāstra,* such as the *Sūrya-siddhānta* and the *Brahmagupta-siddhānta,* contain lists of polar ecliptic latitudes and longitudes for stars.[1] These stars include the *yogatārās* (or "principle stars") of the 28 *nakṣatras* or lunar mansions. The list in the *Sūrya-siddhānta* includes seven additional stars, making a total of 35. These are Agastya, Māgavyādha, Agni, Brahmahādaya, Prajāpati, Apāmvatsa, and Āpas.[2] I will call all these stars "*jyotiṣa* stars" and refer to them by their Sanskrit names.

Given a *jyotiṣa* star, it is natural to try to identify which star it corresponds to in the nomenclature of modern Western astronomy. I shall call this the "modern star" corresponding to the given *jyotiṣa* star. The tenth-century Muslim scholar Alberuni was perhaps the first Westerner to seek such identifications.[3] This effort was continued by scholars such as Colebrooke,[4] Burgess,[5] and Kaye[6] in the nineteenth century and B. Dikshit[7] and David Pingree[8] in the twentieth.

I have adopted the set of identifications in Table 1 for the 35 *jyotiṣa* stars. These identifications are based on the discussion given by Burgess in his translation of the *Sūrya-siddhānta.*[9] They are exactly as given by Burgess with the exception of two or three cases where he did not give a clear choice. In those cases I made an arbitrary choice from the alternatives which he offered.

We can compare a given *jyotiṣa* star position with the corresponding modern star position in the following way: First convert the polar ecliptic latitude and

---

1 Dikshit, S. B., *English Translation of Bharatiya Jyotish Sastra,* Gov. of India Press (Calcutta), 1969, pp. 338–39.

2 Burgess, Ebenezer, *The Sûrya Siddhânta,* Motilal Banarsidass reprint (Delhi), 1989, pp. 245, 252.

3 Sachau, E.C., trans., *Alberuni's India,* Kegan Paul, Trench, Trūbner & Co. (London), 1910, pp. 84– 85.

4 Colebrooke, H. T., 1807, "On the Indian and Arabian Divisions of the Zodiac," *Asiatic Researches,* vol. IX, pp. 323–376.

5 Burgess, pp. 202–52.

6 Kaye, G. R., *Hindu Astronomy,* Cosmo Pub. (New Delhi), 1981, p. 118.

7 Dikshit, p. 345.

8 Pingree, D. and Morrissey, P., "On the Identification of the Yogatārās of the Indian Nakṣatras," *Jour. of the History of Astronomy,* vol. 20, 1989, pp. 99–119.

9 Burgess, ibid.

longitude of the *jyotiṣa* star into right ascension and declination.[10] This gives the position of the star in modern coordinates at its particular epoch. Then apply formulas for precession to obtain the right ascension and declination of the star in the epoch A.D. 2000.[11] These coordinates define the "*jyotiṣa* star position" in modern form. They can be compared with the right ascension and declination of the corresponding modern star listed in a catalog for the epoch A.D. 2000.[12] These latter coordinates define the "modern star position."

A great circle arc can be drawn between the star's modern position and its *jyotiṣa* position. Let D be the length of this arc in degrees. The "error vector" is a vector of length D starting at the modern position and pointing along this arc. According to Burgess, the epoch of the *jyotiṣa* star coordinates is about A.D. 490.[13] The argument for this is that if we precess the *jyotiṣa* star coordinates from various epochs to A.D. 2000, we find that the average of the lengths of the error vectors pointing from the modern star positions to the corresponding *jyotiṣa* star positions are nearly minimal for the epoch A.D. 490. Therefore I have used this epoch when converting *jyotiṣa* star coordinates to the epoch of A.D. 2000.

III Proper Motions of Stars

Many stars are observed to move slowly with respect to the celestial sphere. These motions, called proper motions, vary from star to star. For a typical star, the proper motion amounts to a small fraction of a second of arc per year, and it is directed along a great circle arc on the celestial sphere.

In the star catalog *Sky Catalog 2000.0* (for the epoch A.D. 2000), the present rate of proper motion in both right ascension and declination is listed for each

---

10 I used the following equations to convert polar longitude and latitude (λ,β) into right ascension and declination (α,δ):

$$\alpha = \arctan[\cos(\varepsilon)\sin(\lambda),\cos(\lambda)]$$
$$\delta = \arcsin[\sin(\varepsilon)\sin(\lambda)] + \beta$$

Here arctan[x,y] is the computer arctangent function, which automatically gives an angle lying in the proper quadrant. The quantity ε is the obliquity of the ecliptic. Its current value is about 23.5 degrees. However, I used 24 degrees, since this is the value used in ancient Indian astronomy, and it is the value that would have been used in India to transform star coordinates into polar longitudes and latitudes.

11 For calculations of precession of the equinoxes, I used the standard equations from Green, Robin M., *Spherical Astronomy*, Cambridge Univ. Press (Cambridge), 1985, p. 219 for the epoch of A.D. 2000.

12 I used Hirshfeld, A. and Sinnott, R.W., *Sky Catalogue 2000.0*, Vol. 1, Cambridge Univ. Press (Cambridge), 1982, for all modern star positions and proper motions.

13 Burgess, p. 244.

star. The negatives of these two quantities, expressed in degrees per 10,000 years, define a vector which describes the motion of the star on the celestial sphere as we follow it back in time. I will use the term "reversed proper motion vector" to refer to this vector. (For some stars the distance from the earth and the radial velocity are also listed. These quantities can be used to compute an acceleration or deceleration of the proper motion along the celestial sphere.)

I should emphasize that the right ascension and declination of a star are meaningful only in reference to a particular epoch. This is due to the gradual shifting of these coordinates caused by the precession of the equinoxes. In general, whenever I refer to a star's position, I am referring to its right ascension and declination for the epoch A.D. 2000. This provides a standard way of specifying the position of a star on the celestial sphere. The movement of a star over thousands of years due to proper motion is also expressed in terms of right ascension and declination for A.D. 2000.

Let us suppose, for the sake of argument, that the position of a star was measured long ago with a certain error, and the results of this measurement were passed down to the present as the *jyotiṣa* coordinates for the star. As time went by, the position of the star changed as a result of the star's proper motion. Thus the error vector pointing from the modern star position to the *jyotiṣa* position must consist of two components: one due to proper motion and the other due to inevitable errors in measurement. If the error in measurement is small compared with the shift due to proper motion, then we would expect the modern star position to move towards the *jyotiṣa* position as we follow its movement backward in time.

As we track its motion backwards in time, the star follows a great circle arc that begins at the modern position and extends in the direction of the reversed proper motion vector. Another great circle arc can be drawn between the star's modern position and the *jyotiṣa* position. We shall call the angle A between these two arcs the "angle of approach." This angle lies between 0 and 180 degrees. It is the angle between the reversed proper motion vector and the error vector, as defined above.

As we follow the star's position backward in time, there comes a time when the distance between this position and the *jyotiṣa* position is minimal. Define T to be this time in years before A.D. 2000 and D_{min} to be the minimum distance in degrees. D_{min} can be compared with D, the length of the error vector pointing from the modern star position to the *jyotiṣa* star position. If the angle of approach A is greater than 90 degrees, then the star's position moves further away from the *jyotiṣa* position as we go back in time. In this case, T will be 0 and $D_{min} = D$.

The average value of D is 3.09 degrees for the 35 stars listed in the *Sūrya-siddhānta*, using the star identifications of Burgess. The *Brahmagupta-siddhānta* gives a list of coordinates for the 28 *yogatārās* of the *nakṣatras*.[14] On the whole, these coordinates seem to be more accurate than the coordinates for the 28 *nakṣatras* in the *Sūrya-siddhānta* list. We have therefore created a composite list of star coordinates by adding the seven additional stars found in the *Sūrya-siddhānta* to the *Brahmagupta-siddhānta* list. For this composite list the average value of D is 2.78 degrees. I will use this list of *jyotiṣa* star coordinates in the calculations discussed below.

Figure 1 gives an example of a modern star that moves towards its corresponding *jyotiṣa* star position as we go back in time. The modern star is Capella (Alpha Aurigae), and its present position is marked by the circle labeled "Capella." The position of Capella at successive 10,000 year intervals going into the past is marked by a series of small circles. These approach the large double circle marking the *jyotiṣa* position of Brahmahådaya at approximately 50,000 years ago.

IV Three Hypotheses

The following three hypotheses can be offered regarding possible relationships between *jyotiṣa* star coordinates and proper motions of stars:

1. Null hypothesis. There is no systematic relationship between the proper motions of stars and the error vectors pointing from modern star positions to corresponding *jyotiṣa* star positions. We therefore expect angles of approach from modern star positions to *jyotiṣa* star positions to be more or less random.

2. Historical hypothesis. The *jyotiṣa* star positions were measured in the remote past, and these stars have moved considerably since that time due to proper motions. Therefore, angles of approach tend to be smaller than they should be on the basis of chance expectation. We can learn about the time of measurement of the *jyotiṣa* star positions by computing the times T defined above.

3. Unknown alternative. There is a systematic relationship between the proper motion vectors of stars (going backwards in time) and the error vectors pointing from modern star positions to corresponding *jyotiṣa* star positions. However, the historical interpretation is incorrect, and some other explanation is required to account for this relationship.

14 Dikshit, p. 338–39

V The Null Hypothesis

At first glance, one would expect the null hypothesis to be correct. After all, proper motions of stars are very small, and they were measured only recently using powerful modern telescopes. They should have nothing to do with errors in naked-eye measurements made many centuries ago. However, it turns out that the null hypothesis can be ruled out by a statistical study of the data. I will begin by giving this statistical analysis.

First, one might suppose that error vectors based on *jyotiṣa* star coordinates should point in random directions relative to reversed proper motion vectors. This suggests that the angles of approach A_i (for stars i=1 to 35) should be uniformly distributed between 0 and 180 degrees. The theoretical mean for one angle selected according to the uniform probability distribution on [0,180] is $\mu = 90$ degrees, and the theoretical standard deviation is $\sigma = 180/12^{1/2} = 51.96$ degrees. The theoretical mean and variance for an average of 35 angles selected independently according to this distribution are $\mu' = 90$ and $\sigma' = \sigma/35^{1/2} = 8.78$ degrees, respectively.

If we compute the angles of approach using the star coordinates and star identifications given above, we find an average A_i of 55.78 degrees. This is about 3.9 standard deviations below $\mu' = 90$ degrees.

This finding seems to be contrary to what we might expect from the null hypothesis. However, the directions of the 35 reversed proper motion vectors are distributed nonuniformly, with a bias towards the north. Likewise, the directions of the 35 error vectors are also distributed nonuniformly. Could it be that these nonuniform distributions are responsible for the unexpectedly small angles of approach?

To answer this, it is necessary to take a deeper look at the statistics of the angles of approach. One way to do this is to assign to each star the proper motions of each of the 35 stars. This results in 35×35 = 1225 artificial "stars" which can be used to compute angles of approach. These 1225 combinations have the following properties:

(1) They have the same distributions of error vectors and reversed proper motion vectors as the 35 real stars.

(2) In the 35×35 combinations, each error vector is associated once with each proper motion vector. This erases any particular relationship between error vectors and proper motion vectors that may have existed in the original set of 35 stars.

So if the nonuniform distributions of error and reversed proper motion vectors are responsible for the small angles of approach, then we would also expect the 1225 combinations to yield small angles of approach on the average. The mean angle of approach for the 1225 artificial combinations is $\mu = 82.06$ degrees, and its standard deviation is $\sigma = 50.92$ degrees. This mean is still quite a bit higher than the average of 55.78 degrees that we found for the 35 real stars.

Given the distribution of angles created by the 1225 combinations, how probable is it that the average angle of approach will be as low as 55.78 degrees? The theoretical standard deviation for the average of 35 angles of approach chosen independently according to this distribution is $\sigma' = 50.92/35^{1/2} = 8.61$. Thus 55.78 degrees is 3.05 standard deviations below the mean of $\mu'=82.06$ degrees. This is smaller than 3.9 standard deviations, but it is still a statistically significant deviation from the mean.

Additional evidence against the null hypothesis can be obtained by looking at certain weighted averages of the angles of approach. It turns out that if we use certain weights based on historical studies of the *nakṣatras*, we find that the statistical significance of the correlation between reversed proper motion vectors and error vectors becomes greater. I will explain this after first introducing the idea of a weighted average. A weighted average is given by the following formula:

$$\chi = \sum_{i=1}^{N} W_s A_i \Big/ \sum_{i=1}^{N} W_i \qquad [1]$$

where the W_i's are nonnegative weights.

One set of weights measures the degree of certainty of scholars in the identification of modern stars corresponding to the *yogatārās* of the 28 *nakṣatras*. Six different scholars have given from 1 to 4 different choices for these 28 *yogatārās*.[15] The number Alt_i of choices for *yogatārā* i can be interpreted as a measure of the uncertainty of scholars regarding the identity of the *yogatārā*. Thus 4-Alt_i can be seen as a measure of the scholars' degree of certainty. The average computed using W_i = 4-Alt_i tends to emphasize those *nakṣatras* for which scholars are in agreement on the identification of the *yogatārā*.

Another set of weights represents the degree of success of Alberuni in his attempt to identify the *yogatārās*.[16] These weights, called Alb_i, are defined as follows: $Alb_i = 2$ if Alberuni felt he was successful in identifying the *yogatārā* i; $Alb_i = 1$ in the one case where he seemed to make an ambiguous statement; and $Alb_i = 0$ if Alberuni said he could not identify *yogatārā* i.

A third set of weights, called $Sieu_i$, has to do with correspondences between

---

15 Dikshit, p. 345.

16 Sachau, pp. 84–85.

the Chinese lunar mansions called *sieus* and the *yogatārās* of the *nakṣatras*. Each Chinese *sieu* is a single star. $Sieu_i = 0$ for *nakṣatra* i if Burgess concluded that the corresponding *sieu* does not lie in the *nakṣatra* constellation (which may consist of several stars); $Sieu_i = 1$ if the *sieu* might be in the constellation; $Sieu_i = 2$ if it is in the constellation but is not the *yogatārā*; and $Sieu_i = 3$ if it is the *yogatārā*. The three sets of weights $4\text{-}Alt_i$, Alb_i, and $Sieu_i$, are listed in Table 2.

Table 3 lists the results of calculating average angles of approach using these weights. In line 1 of the table, all 35 stars were used and the weights were $W_i=1$. In this case α is simply the average of the angles of approach for the 35 stars, and the results are as reported above.

In line 2, the average was restricted to the 28 *nakṣatras* and $W_i=1$. For this line, all the calculations are as before, except that 28 stars were used instead of 35. These calculations made use of the set of 28×28 = 784 artificial star combinations created by combining error vectors and proper motion vectors for the 28 *nakṣatras*. In this case the average angle of approach for the 28 *nakṣatras* turns out to be 2.17 standard deviations below the mean calculated for the 28×28 artificial combinations. (This lower value is partly due to the fact that we are using a smaller sample size.)

What happens when we look at more general weighted averages of the A_i's? If 28 A_i's are chosen independently from a distribution with mean μ and standard deviation σ, then the mean of the α given in equation (1) is simply $\mu' = \mu$. The standard deviation of this α is

$$\sigma' = \sigma \times [\, \sum_{i=1}^{N} W_i^2]^{1/2} \;/\; [\, \sum_{i=1}^{N} W_i] \qquad [2]$$

This reduces to the familiar formula $\sigma' = \sigma/N^{1/2}$ in the case where $W_i=1$. Here N=28.

Using this formula, we can compute the number of standard deviations separating the weighted average of the angles of approach and the mean calculated for the 28×28 artificial combinations. We find that for the weights $W_i = 4\text{-}Alt_i$ and Alb_i, the statistical significance of the deviation from the mean is greater than in the case where $W_i=1$. In place of 2.17 standard deviations, we get 2.52 and 2.41, respectively.

This means that by emphasizing *nakṣatras* for which scholars were confident of their identification of the *yogatārā*, we obtain a more significant correlation between reversed proper motion vectors and vectors indicating errors in *jyotiṣa* star coordinates. Likewise, the correlation is more significant when we give emphasis to stars that Alberuni felt he could identify. A natural interpretation of this is that with these weights we are excluding (or de-emphasizing) *nakṣatras* which may have been falsely identified. By the null hypothesis, the angles of approach for falsely identified stars should not be significantly different from these for correctly identified stars (since all angles of approach should be more or less random). But this is not what we find.

Using the weights $Sieu_i$, we find that the weighted average is 2.56 standard deviations below the mean. In this case it turns out that we acquire a more significant correlation between error vectors and proper motions when we look at *nakṣatras* that are thought to correspond to Chinese *sieus*. As with $4\text{-}Alt_i$ and Alb_i, this set of weights may tend to select stars that are more securely identified. By the null hypothesis, these *nakṣatras* should not tend to have lower angles of approach than other *nakṣatras*, but it appears that they do.

VI The Historical Hypothesis

Since the null hypothesis does not hold up, let us turn to the historical hypothesis. If this hypothesis is correct, then the times of closest approach for stars with small angles of approach should be estimates of how long ago the *jyotiṣa* coordinates of these stars were measured. The following string of numbers is a histogram of the times of closest approach for those stars out of the 35 that have angles of approach less than 45 degrees:

410224000110001

The first number gives the number of stars with times between 0 and 5,000 years ago. The remaining numbers give the numbers of stars with times falling in successive intervals of 10,000 years. We can see that there is one peak corresponding to the interval between 0 and 5,000 years ago and another peak corresponding to the broad interval between 25,000 to 55,000 years ago.

For the moment, let us disregard the extremely large ages represented by these peaks and try to evaluate their statistical significance. It turns out that it is easiest to do this if we restrict our attention to the 28 *nakṣatras*. For these stars, the histogram is:

410213000000000

We can create a weighted histogram using weights W_i by letting each number in the histogram be the sum of the W_i's for the stars falling in that number's time interval. If we do this for $W_i = Sieu_i$, we get:

000536000000000

It turns out that the five stars making up the more recent peak all have $Sieu_i$ = 0 (meaning that none are Chinese *sieus*), and thus this peak does not appear in the weighted histogram. These stars are Punarvasu (Beta Geminorum), Maghā (Alpha Leonis), Svātī (Alpha Bootis), Abhijit (Alpha Lyrae), and Śravaṇa (Alpha Aquilae). They are all prominent stars with high proper motions, and they all fall in the set of nine *nakṣatras* with magnitudes less than 1.5. Note that the probability that five randomly chosen *nakṣatras* will fall in this group of nine is less than $(9/28)^5$. (In fact, it is 1/780.)

Given the null hypothesis, this histogram peak made of prominent stars must simply be a product of chance. However, the historical hypothesis provides a simple explanation for it. We can argue that since these stars are fast moving and prominent to the eye, their positions must have been measured relatively recently. Otherwise, they would have moved so far from their recorded positions that those positions would have been obviously in error. Burgess argues that the observed correspondences between *nakṣatras* and *sieus* must be due to cultural contact between Indian and Chinese astronomers. We can hypothesize that the stars in the recent peak do not correspond to *sieus* because they were added to the *nakṣatra* system after this period of close cultural contact. Although this explanation is certainly speculative, it does provide possible reasons for a pattern which, according to the null hypothesis, must be entirely due to chance.

Now let us consider the histogram peak in the interval from 25,000–55,000 years ago. This peak is a concentration of ages around a central point of 40,000 years. The question is: Is this concentration of ages statistically significant, or is it just a product of chance?

We can answer this question by taking advantage of the fact that $Sieu_i$ cancels the recent peak in the histogram and leaves the 25,000–55,000 year peak. This gives us a criterion for singling out the older peak which is independent of our findings regarding proper motions and error vectors. (It is independent because $Sieu_i$ depends only on Burgess's study.) Let $D4_i$ be the distance in degrees between the ith *jyotiṣa* star position and the position occupied by the corresponding modern star 40,000 years ago. If the 25,000–55,000 year peak is significant, then the weighted average of the $D4_i$'s using $W_i = Sieu_i$ should be unexpectedly small.

This turns out to be true. The weighted average of the $D4_i$'s using $Sieu_i$ as weights is 1.47 standard deviations below the mean for this average calculated using the 784 artificial stars. If we use the weights $Sieu_i \times Alb_i$ which reflect both the correspondence with Chinese *sieus* and Alberuni's star identifications, the corresponding figure becomes 1.7 standard deviations. Thus we can conclude that the 25,000–55,000 year peak has some statistical significance, but it is not highly significant.

However, there is additional evidence suggesting that this peak should be taken seriously. If we observe the individual diagrams showing the relation between modern star positions and *jyotiṣa* position, we can find several instances in which the *yogatārā* identified by Burgess does not closely approach the *jyotiṣa* position as we go back in time over an 80,000 year period. However, in some of these instances there is another star in the *nakṣatra* constellation that does closely approach the *jyotiṣa* position in this time period. Table 4 lists these new star identifications for Bharaṇī, Māgaśīrṣa, Puṣya, Pūrvaphalgunī, and Mūla.

If we adopt these five new *yogatārā* identifications, what effect does this have on the histogram of ages? Does it tend to disperse the two age peaks that we have already noticed, or does it accentuate them? The age histograms for all 35 stars and for the 28 *nakṣatras* become:

410226110110001

410215110000000

If we compare these with the corresponding histograms for the pure Burgess star identifications, we find that the two age peaks remain, and that the number of stars in the 45,000 –55,000 year range has become greater. With these modified star identifications, we can again calculate the weighted average of the $D4_i$'s using the $Sieu_i$'s as weights. This weighted average is now 2.0 standard deviations below the mean based on the 784 artificial stars. If we use the $Sieu_i \times Alb_i$'s as weights, this figure becomes 2.3 standard deviations.

These results indicate that the 25,000 –55,000 year peak has become more prominent as a result of the new star identifications. These identifications were made on the basis that the chosen star should closely approach the *jyotiṣa* star position as one goes back in time. They were not specifically chosen so that it would approach the *jyotiṣa* position in the interval of 25,000–55,000 years ago. Yet the ages of closest approach for the five new identifications are 40.7, 53.5, 73.9, 64.4, and 48.2 thousand years, respectively. Thus they do fall on the high side of this peak.

Figure 2 illustrates the choice of the new star identification in the case of Müla. Burgess identifies the *yogatārā* of this *nakṣatra* as Lambda Scorpionis, but several other authorities disagree. We can see from the figure that Lambda Scorpionis barely moves over a 50,000 year period. However, Epsilon Scorpionis moves almost directly towards the *jyotiṣa* position of Müla as we go back in time. Thus by visual inspection, it is a natural replacement for Lambda.

VII Summary Discussion of the Three Hypotheses

According to the historical hypothesis, the 25,000 –55,000 year peak represents a period of astronomical measurement that took place roughly 25,000–55,000 years ago. Of course, the obvious objection to this is that at this time civilization didn't exist. Astronomy and the transmission of astronomical knowledge from generation to generation are thought to be impossible in this period.

It is generally accepted that human beings of modern type have existed for at least 40,000 years. But agriculture and settled village life are thought to have arisen only 7,000–10,000 years ago.[17] Early *Homo sapiens sapiens* presumably had the same talents and capabilities as modern humans, and one wonders why they waited 30,000–33,000 thousand years before starting the explosive development of civilized life. Could it be that civilization arose more gradually over a longer period of time?

Many ancient peoples apparently thought so. For example, fragments of the writings of Manetho, an Egyptian historian dated to the 3rd century B.C., give a period of some 30,000 years to early Egyptian civilization.[18] Berossus, a Babylonian historian of the same period, assigned 432,000 years to early Babylonian history,[19] and a similar time span is found in the king lists of Sumeria.[20] The period of 432,000 years is, of course, reminiscent of the Indian yuga cycles, even though these are regarded by some indologists as a recent invention by astronomers.[21]

It is easy to dismiss these traditions as sheer mythology, but it is possible that they may contain some truth. Perhaps sophisticated arts and sciences go back further than is generally believed.

Nonetheless, we may choose to reject the historical hypothesis. If we also reject the null hypothesis, then we are left with hypothesis three: There is a significant relationship between proper motions of stars and errors in *jyotiṣa* star positions, but this has nothing to do with the slow drift of stars after an ancient period of astronomical measurement. Unfortunately, since proper motions of stars are exceedingly slow, it is hard to see how their effects could become manifest over short periods of time to pre-telescopic astronomers. Long periods of time would seem to be necessary, and this suggests that some form of historical explanation is required.

17 Gowlett, John, *Ascent to Civilization,* Collins (London), 1984, pp. 120, 156.

18 Waddell, W. G., *Manetho,* Harvard Univ. Press (Cambridge), 1964.

19 Burstein, Stanley M., *The Babyloniaca of Berossus,* Undena Publications (Malibu), 1978.

20 Lambert, W. G. and Millard, A. R., *Atrahasis: The Babylonian Story of the Flood,* Clarendon Press (Oxford), 1969, p. 17.

21 Sethna, K. D., *Ancient India in a New Light,* Aditya Prakashan (New Delhi), 1989, p. 133.

I will therefore close by summing up the evidence presented in this paper that goes against the null hypothesis. This evidence can be summed up as follows:

1. The data for the 35 pairs of modern and *jyotiṣa* stars can be used to create an artificial distribution which has the same distributions of proper motions and error vectors as the 35 pairs, but eliminates all correlations between proper motions and error vectors. The null hypothesis predicts that the average angle of approach α for the 35 stars should be about the same as the average angle of approach μ for this artificial distribution. But actually α is significantly lower than μ, as the historical hypothesis predicts. This also holds true if we restrict our analysis to the 28 *nakṣatras*.

2. We can formulate three sets of weighting factors based on historical studies of the *nakṣatras* having nothing to do with proper motions of stars. These weights have to do with Chinese *sieus*, the *nakṣatra* identifications of Alberuni, and *nakṣatra* identifications by a number of other scholars. When the average angle of approach α is calculated using these weights, it becomes lower and its statistical significance increases. According to the null hypothesis, this should not happen, but it can be readily explained in terms of the historical hypothesis.

3. The histogram of age estimates shows two pronounced peaks. The recent peak, corresponding to the period between 0 and 5,000 years ago, is made up of bright, prominent stars, and none of these stars correspond to Chinese *sieus*. According to the null hypothesis, it is quite improbable for this to happen. However, it is amenable to a simple historical explanation.

4. We can evaluate the statistical significance of the older histogram peak by looking at distances $D4_i$ of modern stars from *jyotiṣa* star positions 40,000 years ago, at a time corresponding to the middle of this peak. If we use the weights $Sieu_i$, which eliminate the recent peak, we find that these distances are significantly lower than would be expected by the null hypothesis. The level of significance is 1.5 standard deviations using $Sieu_i$ and 1.7 standard deviations using $Sieu_i \times Alb_i$.

5. If we visually examine graphs of star movement for the 28 *nakṣatras*, we can find five cases where the *yogatārā* chosen by Burgess does not move towards the *jyotiṣa* star position as we go back in time, but there is another star in the *nakṣatra* constellation which does this. All of the results reported in points 1–4 were based on Burgess's star identifications. If we substitute the five new star identifications, we would naturally expect the average angles of approach to become smaller. This is because the new identifications were chosen with this in mind. However, we wouldn't necessarily expect the age estimates for these new identifications to reinforce the 25,000–55,000 year peak. There are not enough stars in the *nakṣatra* constellations to give us the opportunity to

deliberately make such identifications. Nonetheless, the five new identifications all fall close to the 25,000 –55,000 year peak (and shift its center a bit closer to 55,000 years). This is also unexpected, given the null hypothesis.

These five points add up to a strong case against the null hypothesis, and they give some direct support to the historical hypothesis. I would suggest that it might therefore be worthwhile to seek further evidence of high cultural achievements in time periods going back tens of thousands of years.

FIGURES

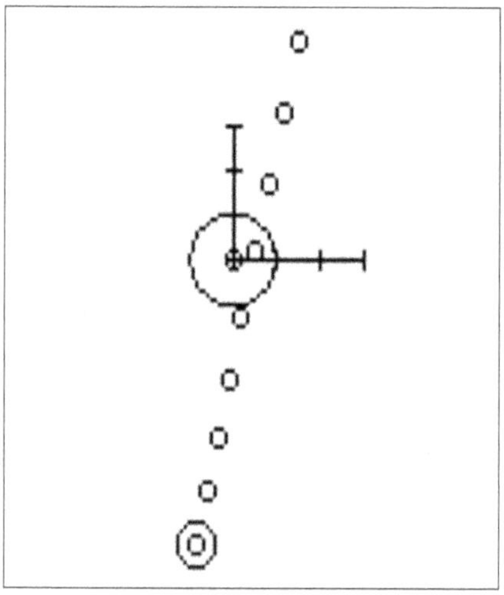

Figure 1. The large circle marks the *jyotiṣa* position of Brahmahṛdaya (see line 32 of Table 1). Burgess identifies this star as Capella (Alpha Aurigae). The figure shows that Capella (double-ringed) was very close to the position of Brahmahṛdaya about 50,000 years ago. Its distance will be 6.487 degrees in A.D. 2000, and it reached a minimum value of 0.453 degrees in 48,164 B.E. (B.E. means before the epoch of A.D. 2000.)

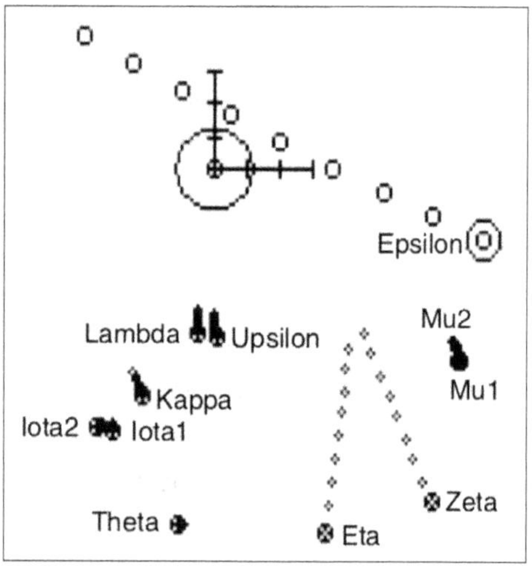

Figure 2. The large circle marks the *jyotiṣa* position of Mūla (see line 19 of Table 1). Colebrooke and Burgess agree that this *nakṣatra* should consist of the stars Upsilon Lambda Kappa Iota Theta Eta Zeta Mu Epsilon Scorpionis that form the tail of the scorpion.

The *Sūrya-siddhānta* says that the *yogatārā* is the eastern star of the group, and this is Iota. However, Colebrooke selects Upsilon and Burgress selects Lambda. This is done on the basis of position and agreement with the Arabic manzil ash-Shaulan, which consists of Upsilon and Lambda.

It turns out that Upsilon and Lambda barely move over a 50,000 year period. However, the figure shows that the star Epsilon reached a minimum distance of 1.667 degrees from the position of Mūla in 48,223 B.E. Therefore I have chosen Epsilon in Table 4 as a replacement for Burgess's choice of Lambda.

TABLES

TABLE 1

No.	Jyotiṣa star	Modern star	D	D_{min}	D4/D	A
1	Aśvinī	Beta Arietis	1.27	.58	.59	27.12
2	Bharaṇī	35 Arietis	1.29	1.08	.95	57.13
3	Kṛttikā	Eta Tauri	.61	.09	.19	8.75
4	Rohiṇī	Alpha Tauri	1.65	.24	.53	8.52
5	Mṛgaśīrṣa	Lambda Orionis	4.40	2.87	.99	40.76
6	Ārdrā	Alpha Orionis	6.26	6.25	1.00	86.84
7	Punarvasu	Beta Geminorum	.69	.31	9.33	26.55
8	Puṣya	Delta Cancri	1.75	1.58	1.41	64.76
9	Āśleṣā	Epsilon Hydrae	5.23	5.21	1.06	85.34
10	Maghā	Alpha Leonis	.21	0.15	12.32	43.31
11	Pūrvaphalgunī	Delta Leonis	3.77	3.77	1.51	151.18
12	Uttaraphalgunī	Beta Leonis	.41	.41	14.62	145.01
13	Hasta	Delta Corvi	2.84	.49	.17	9.89
14	Citrā	Alpha Virgini	.94	.58	.64	38.09
15	Svātī	Alpha Bootis	2.85	.43	7.25	8.64
16	Viśakhā	24 Iota Librae	2.59	2.23	.90	59.60
17	Anurādhā	Delta Scorpii	3.05	3.03	.99	82.29
18	Jyeṣṭhā	Alpha Scorpii	1.50	1.20	.90	53.07
19	Mūla	Lambda Scorpii	5.02	.33	.94	3.83
20	Pūrvaṣāḍhā	Delta Sagittari	1.25	1.25	1.22	111.21
21	Uttaraṣāḍhā	Sigma Sagittari	2.31	2.31	1.23	146.56
22	Abhijit	Alpha Lyri	1.68	.99	1.46	36.02
23	Śravaṇa	Alpha Aquilae	.71	.38	7.72	32.23
24	Dhaniṣṭhā	Beta Delphini	3.20	2.96	.92	67.58
25	Śatabhiṣaj	Lambda Aquarii	67	.43	65	40.10
26	Pūrvabhādrapadā	Alpha Pegasi	3.46	3.21	.94	68.34
27	Uttarabhādrapadā	Alpha Andromedo	5.78	5.78	1.09	92.21
28	Revatī	Zeta Piscium	1.18	1.18	2.15	124.16
29	Agastya	Alpha Carinae	.85	.11	.55	7.45
30	Mṛgavyādha	Alpha Can. Maj.	1.12	.99	11.13	62.61
31	Agni	Beta Tauri	7.92	7.12	.92	63.97
32	Brahmahṛdaya	Alpha Aurigae	6.49	.45	.20	4.01
33	Prajāpati	Delta Aurigae	6.57	2.50	77	22.40
34	Apāmvatsa	Theta Virginis	1.84	1.16	0.79	39.02
35	Āpas	Delta Virginis	5.91	3.29	0.56	33.83

Table 1. Identification of modern stars corresponding to stars listed in the *Sūrya-siddhānta*. These identifications were given by Burgess. The quantities D, D_{min}, D4/D, and A were calculated for this set of identifications. These quantities are defined in the text.

TABLE 2

Nakṣatra	Modern star	4-Alt	Alb	Sieu	Sieu × Alb
Aśvinī	Beta Arietis	2	2	3	6
Bharaṇī	35 Arietis	1	1	3	3
Kṛttikā	Eta Tauri	3	2	3	6
Rohiṇī	Alpha Tauri	3	2	2	4
Mṛgaśīrṣa	Lambda Orionis	2	2	3	6
Ārdrā	Alpha Orionis	1	0	0	0
Punarvasu	Beta Geminorum	3	2	0	0
Puṣya	Delta Cancri	3	2	3	6
Āśleṣā	Epsilon Hydrae	0	0	2	0
Maghā	Alpha Leonis	3	2	0	0
Pūrvaphalgunī	Delta Leonis	2	2	0	0
Uttaraphalgunī	Beta Leonis	3	2	0	0
Hasta	Delta Corvi	2	2	3	6
Citrā	Alpha Virgini	3	2	3	6
Svātī	Alpha Bootis	3	2	0	0
Viśakhā	24 Iota Librae	1	0	2	0
Anurādhā	Delta Scorpii	2	2	2	4
Jyeṣṭhā	Alpha Scorpii	3	2	2	4
Mūla	Lambda Scorpii	0	2	2	4
Pūrvaṣāḍhā	Delta Sagittari	2	2	1	2
Uttaraṣāḍhā	Sigma Sagittari	0	2	1	2
Abhijit	Alpha Lyri	3	2	0	0
Śravaṇa	Alpha Aquilae	3	2	0	0
Dhaniṣṭhā	Beta Delphini	2	0	0	0
Śatabhiṣaj	Lambda Aquarii	3	0	0	0
Pūrvabhādrapadā	Alpha Pegasi	3	0	3	0
Uttarabhādrapadā	Alpha Andromedo	2	0	2	0
Revatī	Zeta Piscium	2	0	0	0

Table 2. Weighting factors. These weights represent the degree of certainty of scholars in the identification of the *yogatārās* of the 28 *nakṣatras* (4-Alt and Alb) and the degree of correspondence between Chinese lunar mansions and *nakṣatras* (Sieu). The weight Sieu×Alb is simply the product of the weights Sieu and Alb.

TABLE 3

	Weights	α	Mean	S.D.	No. of S.D.s
1	35 Ones	55.78	82.06	8.61	-3.05
2	28 Ones	61.40	82.43	9.67	-2.17
3	4-Alt_i	55.61	82.43	10.65	-2.52
4	Alb_i	54.69	82.43	11.51	-2.41
5	$Sieu_i$	49.30	82.43	12.92	-2.56
6	$Sieu_i \times Alb_i$	40.92	82.43	14.95	-2.78

Table 3. Weighted averages of the angles of approach. Here α is the weighted average of the A_i's using the indic ated weights. In line 1 the average is for 35 stars with unit weights. In the remaining lines the average is limited to the 28 *nakṣatras*. The theoretical mean and standard deviation for 35×35 or 28×28 artificial stars are listed along with the number of standard deviations between α and the mean.

TABLE 4

No.	Jyotiṣa star	Modern star	D	D_{min}	D4/D	A
1	Bharani	41 Arietis	2.69	2.25	.84	56.67
2	Mrgasirsa	Phi2 Orionis	5.16	.82	.31	9.20
3	Pusya	Theta Cancri	1.82	.50	.53	16.05
4	Pürvaphalguni	Theta Leonis	1.88	.02	.38	.54
5	Mula	Epsilon Scorpii	8.82	1.67	.25	10.93

Table 4. These star identifications are replacements for the corresponding identifications of Burgess in Table 1. Each replacement star belongs to the same *nakṣatra* constellation as Burgess's *yogatārā*, but it shows a stronger tendency to approach the *jyotiṣa* position as we go back in time.

CHAPTER 6

Planetary Diameters in the *Sūrya-siddhānta*

Originally published in:
Journal of Scientific Exploration
Volume 11, Number 2 (1997), pp. 193–200

Abstract

This paper discusses a rule given in the Indian astronomical text *Sūrya-siddhānta* for computing the angular diameters of the planets. By combining these angular diameters with the circumferences of the planetary orbits listed in this text, it is possible to compute the diameters of the planets. When these computations are carried out, the results agree surprisingly well with modern astronomical data. Several possible explanations for this are discussed, and it is hypothesized that the angular diameter rule in the *Sūrya-siddhānta* may be based on advanced astronomical knowledge that was developed in ancient times but has now been largely forgotten.

Introduction

In chapter 7 of the *Sūrya-siddhānta* (Burgess, 1989), the 13th verse gives the following rule for calculating the apparent diameters of the planets Mars, Saturn, Mercury, Jupiter, and Venus:

> 7.13. The diameters upon the moon's orbit of Mars, Saturn, Mercury, and Jupiter, are declared to be thirty, increased successively by half the half; that of Venus is sixty.

The meaning is as follows: The diameters are measured in a unit of distance called the *yojana*, which in the *Sūrya-siddhānta* is about five miles. The phrase "upon the moon's orbit" means that the planets look from our vantage point as though they were globes of the indicated diameters situated at the distance of the moon. (Our vantage point is ideally the center of the earth.) Half the half of 30 is 7.5. Thus the verse says that the diameters "upon the moon's orbit" of the indicated planets are given by 30, 37.5, 45, 52.5, and 60 *yojanas*, respectively.

The next verse uses this information to compute the angular diameters of the planets. This computation takes into account the variable distance of the planets from the earth, but for the purposes of this paper it is enough to consider the angular diameters at mean planetary distances. The diameters upon the moon's orbit were given for the planets at these mean distances from the earth. The *Sūrya-siddhānta* says that there are 15 *yojanas* per minute of arc at the distance of the moon (giving 324,000 *yojanas* as the circumference of the moon's orbit). Thus the mean angular diameters of the planets can be computed by dividing the diameters upon the moon's orbit by 15. Table 1 gives the results of this computation and lists other estimates of planetary angular diameters for comparison.

TABLE 1

Angular Diameters of Planets in Minutes of Arc

Planet	*Surya-Siddhanta*	Ptolemy	Tycho Brahe	Modern Minimum	Modern Maximum
Mars	2.0	1.57	1.67	0.058	0.392
Saturn	2.5	1.74	1.83	0.249	0.344
Mercury	3.0	2.09	2.17	0.076	0.166
Jupiter	3.5	2.61	2.75	0.507	0.827
Venus	4.0	3.13	3.25	0.159	1.050

The modern angular diameters are for the greatest and least distances of the planets from the earth.

The *Sūrya-siddhānta* figures are roughly the same size as the planetary angular diameters reported by the 2nd century Alexandrian astronomer Claudius Ptolemy in his book Planetary Hypotheses. Ptolemy attributed his angular diameters to the Greek astronomer Hipparchus, but he did not say how they were measured. According to the historian of astronomy Noel Swerdlow (1968), no earlier reports of planetary angular diameters are known, and Ptolemy's angular diameters were reproduced without change by later GrecoRoman, Islamic, and European astronomers up until the rise of modem astronomy in the days of Galileo, Kepler, and Tycho Brahe.

Brahe's figures were obtained by sighting through calibrated pinholes by the naked eye. They are very similar to Ptolemy's, and they are clearly much larger than the angular diameters measured in more recent times by means of telescopes (Burgess, 1989). It is well known that a small, distant light source looks larger to the naked eye than it really is. This phenomenon makes it likely that angular diameters of planets would inevitably have been overestimated by astronomers before the age of the telescope.

It has been argued that Indian astronomy was heavily influenced by Hellenistic astronomy between the second and fifth centuries A. D. (Pingree, 1976). This suggests that the angular diameters given in the *Sūrya-siddhānta* may have been based on Ptolemy's angular diameters. Indeed, Ptolemy's figures are very close to 94/(60 -7.5n), where n+1 is the line number in Table 1. The corresponding *Sūrya-siddhānta* figures are given by (30 + 7.5n)/15.

Whether or not this indicates an Indian adaptation of Greek material, the angular diameters from *Sūrya-siddhānta* have an important property that the Ptolemaic angular diameters lack. To see this, it is first necessary to examine the sizes of the planetary orbits, as given in *Sūrya-siddhānta*.

Orbital Dimensions in the *Sūrya-siddhānta*

Verses 12.85–90 of the *Sūrya-siddhānta* give the circumferences of the planetary orbits in *yojanas*, and these figures are reproduced in Table 2. The orbits are represented as simple circles centered on the earth, and their circumferences are proportional to the mean orbital periods of the planets. For Mercury and Venus, the mean planetary position is the same as the position of the sun, and thus the orbital circumferences in the table are the same for Mercury, Venus, and the sun. For Mars, Jupiter, and Saturn, the mean position corresponds to the average motion of the planet in its heliocentric orbit.

TABLE 2

Geocentric Orbital Circumferences

Planet	*Surya-Siddhanta* Orbital Circumference (yojanas)	Ptolemy
Moon	324,000	258,000
Mercury	4,332,000	3,447,000
Venus	4,332,000	3,447,000

Planet	*Surya-Siddhanta* Orbital Circumference (yojanas)	Ptolemy
Sun	4,331,500	3,447,000
Mars	8,147,000	6,483,000
Jupiter	51,376,000	40,884,000
Saturn	127,668,000	101,595,000

As given in texts 12.85–90 of the *Sūrya-siddhānta*. The orbital radii are computed from these circumferences using 5 miles per *yojana*.

Verse 1.59 of the *Sūrya-siddhānta* gives the diameter of the earth as 1,600 *yojanas*. Several scholars have argued that the *yojana* in the *Sūrya-siddhānta* is about 5 miles, thereby bringing the earth's diameter to the realistic value of 5×1600 = 8,000 miles. Examples are Sarma (1956), Burgess (1989), and Dikshit (1969).

Different standards were adopted for the *yojana* by different medieval Indian astronomers. This was noted by the astronomer Paramesvara (1380–1450 A.D.), who said:

> What is given by Aryabhata as the measure of the earth and the distances [of the Planets from it], etc., is given as more than one and a half times by other [astronomers]; this is due to the difference in the measure of the *yojana* [adopted by them] (Sarma, 1956).

Verse 4.1 of the *Sūrya-siddhānta* gives the diameters of the sun and moon as 6,500 and 480 *yojanas*, respectively. Given 5 miles per *yojana*, the resulting lunar diameter of 5×480 = 2,400 miles is about 11% higher than the modem value. The corresponding earth-moon distance of about 258,000 miles (listed in Table 2) is high by 8.3%. However, the sun's diameter comes to 5×6500 = 32,500 miles, which is far too small.

It is easy to see why the diameter of the moon should be reasonably accurate. The dimensions of the moon and its orbit were well known in ancient times. For example, the lunar diameter given by Ptolemy in his Planetary Hypotheses falls within about 7% of the modem value, if we convert his earth-diameters into miles using the modem diameter of the earth (Swerdlow, 1968).

It is also easy to see why the diameter for the sun is too small. Ancient astronomers tended to greatly underestimate the earth-sun distance, and Table 2 shows that this also happened in the *Sūrya-siddhānta*. The angular diameter of the sun is easily seen to be about the same as that of the moon – about 1/2 degree. This angular diameter, combined with a small earth-sun distance, leads inevitably to a small estimate for the diameter of the sun. Ptolemy's solar diameter figure is similar to the *Sūrya-siddhānta's*.

Computing Planetary Diameters

What about the planets? Ptolemy listed wildly inaccurate diameters for Mercury, Venus, Mars, Jupiter, and Saturn in his *Planetary Hypotheses.* To see what the *Sūrya-siddhānta* says about the diameters of these planets, we should multiply the orbital radii in Table 2 by the angular diameters (converted to radians) in Table 1. This is done in Table 3.

TABLE 3

Planetary Diameters in Miles

Planet	Modern Diameter	*Surya-Siddhanta* Diameter	% Error
Mercury	3032	3008	–1
Venus	7523	4011	–47
Mars	4218	3772	–11
Jupiter	88748	41624	–53
Saturn	74580	73882	–1

Computed using the *Sūrya-siddhānta* orbital radii from Table 2 and angular diameters from Table 1. The error percentages compare the *Sūrya-siddhānta* diameters with the corresponding modern planetary diameters.

Note that even though the angular diameters are too large, and the orbital radii are too small, the calculated diameters are close to modem values for Mercury, Mars, and Saturn. For Venus and Jupiter, they are too small by about 50%. One might argue that this balancing is due to pure chance. However, since the balancing works for five distinct cases, it is worthwhile to estimate just how probable it is.

This probability can be evaluated by setting up a model in which diameters are chosen at random. One can then check to see if the observed correlation between modem and *Sūrya-siddhānta* diameters is likely to show up in this model. Of course, it is difficult to propose a realistic probabilistic model of how ancient people would have generated astronomical data. But it is possible to set up a simple model in which it is assumed that all planetary diameters, ancient and modern, are given by positive random numbers. It is easy to show that the observed correlation between modern and *Sūrya-siddhānta* diameters is highly unlikely to arise by chance, according to this model. This is discussed in the appendix.

If the observed correlation did not happen by chance, then perhaps it happened by design. One hypothesis is that at some time in the past, ancient astronomers possessed realistic values for the diameters of the planets. They might have acquired this knowledge during a forgotten period in which astronomy reached a high level of sophistication. Later on, much of this knowledge was lost, but fragmentary remnants were preserved and eventually incorporated into texts such as the *Sūrya-siddhānta.* In particular, the real diameters of the planets were later combined with erroneous orbital circumferences to compute the diameters "upon the moon" given in verse 7.13. These figures were then accepted because they gave realistic values for the angular diameters of the planets as seen by the naked eye.

This hypothesis is supported by the fact that the *Sūrya-siddhānta* diameters of Jupiter and Venus in Table 3 are almost exactly half of the corresponding modem diameters. If we multiply these *Sūrya-siddhānta* diameters by 2, we get 83248 miles for Jupiter and 8022 miles for Venus. These figures differ from the corresponding modem values by -6% and +7%. Given this correction, all five planets have an error of 11% or less. (The root-mean-square error comes to 6.3%.) One can argue that the *Sūrya-siddhānta* diameters for Jupiter and Venus were actually the radii for these planets, and somehow they were accepted as diameters by mistake. Or radii might have been deliberately used instead of diameters in order to allow for the simple rule of 30+7.5n used in verse 7.13. This is consistent with the fact that such verses were intended as memory aids and brevity was considered to be a virtue.

Alternative Explanations

Of course, it could be argued that this is just number jugglery, and by juggling numbers one can create false correlations. But let us review the steps taken thus far. The angular diameters in Table 1 were given by the text of the *Sūrya-siddhānta.* The orbital radii of Table 2 were computed from *Sūrya-siddhānta* orbital circumferences using the conversion factor of 5 miles per *yojana.* This factor is based on the *Sūrya-siddhānta's* diameter for the earth, and it has been discussed by other authors. There is no scope for juggling numbers here.

The only proposed adjustment of the numbers is the doubling of the *Sūrya-siddhānta* diameters of Jupiter and Venus. Since the *Sūrya-siddhānta* numbers can be so easily brought into line with modem data, it may be that they have a genuine relationship with this data. One possible explanation is that verse 7.13 may have been written recently, using modern planetary data, and falsely interpolated into the text. But this is ruled out by the fact that there is a manuscript of the *Sūrya-siddhānta* that scholars date to the year 1431 A. D. (Shukla, 1957). This manuscript includes a commentary by Paramesvara, who died in 1450 A. D., and thus it definitely dates back to the 15th century. Verse 7.13 is present in this manuscript, and it agrees with the Burgess translation quoted above. The commentary explains the verse point by point, and thus it confirms that the verse was present in the manuscript in the same form in which it appears today.

In 15th century Europe, the prevailing ideas concerning the sizes of the planets came from medieval Islamic astronomers who were following the teachings of Ptolemy. The first telescopic observations of planets were made by Galileo in 1609–10 (Drake, 1976). As late as 1631, Pierre Gassendi of Paris was shocked when his telescopic observation of a transit of Mercury across the sun revealed that its angular diameter was much smaller than he had believed possible (Van Helden, 1976). It is clear that the information on planetary diameters in the *Sūrya-siddhānta* antedates the development of modem knowledge of these diameters.

It is also clear that Hellenistic astronomers did not have accurate diameters for the planets. Ptolemy computed planetary diameters from his angular diameters and his estimates of planetary distances, and these were reproduced without significant change by European and Islamic astronomers for centuries (Swerdlow, 1968). However, his figures disagree strongly both with modem data and with the diameters computed from *Sūrya-siddhānta* in Table 3.

Deriving the *Sūrya-siddhānta* Rule

If we hypothesize that verse 7.13 incorporates knowledge of the actual diameters of the planets, then one natural question is this: If one started with the modem diameters of the planets and the *Sūrya-siddhānta* orbital circumferences, could one arrive at the rule given in this verse? We can answer this question by computing planetary diameters "upon the moon's orbit" as follows: For each planet, multiply its modem diameter, converted to *yojanas*, by the ratio between the orbital circumferences of the moon and the given planet, as listed in Table 2. Here we use the radius in place of the diameter for Jupiter and Venus. The resulting values are listed in the leftmost column of Table 4.

TABLE 4

Deriving Verse 7.13 from Modern Data

Planet	Modern Projection	Least squares fit	Angular Diameter
Mars	33.6	33.1	2.2
Saturn	37.9	39.4	2.6
Mercury	45.4	45.8	3.0
Jupiter	56.0	52.2	3.5
Venus	56.3	58.5	3.9

The idea behind the rule in verse 7.13 is to arrange the planets so that the diameters on the moon's orbit are in increasing order and then approximate them by a simple arithmetic progression. We can see from Table 4 that the order of the planets used in this rule does put the computed diameters "on the moon's orbit" in increasing order. One can approximate them by an arithmetic progression of the form an+b either by trial and error or by using an optimization method such as least squares. I did this by least squares and got a= 6.356 and b= 33.089. This arithmetic progression is listed in the middle column of Table 4.

In the leftmost column, modern planetary diameters are projected to the orbit of the moon, assuming the planetary orbits given in *Sūrya-siddhānta*. The projected diameters are expressed in *yojanas* (and radii are used in place of diameters for Jupiter and Venus). In the middle column, these projected diameters are fit to an arithmetic progression using least squares. The angular diameters in the rightmost column are obtained by dividing the figures in the middle column by 15 *yojanas* per minute of arc.

One could arrive at the rule in verse 7.13 by observing that 33.1 is about 30, 45.8 is about 45, and 58.5 is about 60. Or one could compute the angular diameters listed in the rightmost column of Table 4 by dividing the numbers in the arithmetic progression by 15. It is plausible that someone looking for a simple rule might round off these angular diameters to the *Sūrya-siddhānta* series of 2,2.5, 3, 3.5, 4.

Thus it is possible to derive the rule in verse 7.13 from modem values for the diameters of the planets.

Conclusion

In summary, verses 7.13 and 12.85–90 of the *Sūrya-siddhānta* contain information regarding the true diameters of the five planets Mercury, Venus, Mars, Jupiter, and Saturn. This information enables us to compute the diameters of three of these planets with errors of 11% or less. If the computed figures for Jupiter and Venus are interpreted as their radii rather than their diameters, then these radii are in error by about 6% and 7%, respectively. This may not be due to mere coincidence. Rather, it may indicate that accurate knowledge of planetary diameters was possessed by ancient astronomers and used in the composition either of the *Sūrya-siddhānta* or of some earlier astronomical text on which it was based. It is not apparent how such knowledge may have been obtained, but we should be on the alert for other possible examples.

References

Burgess, Ebenezer, trans. (1989). *The Sūrya Siddhānta*. Gangooly, P., ed., Delhi: Motilal Banarsidass.

Dikshit, S. B. (1969). *English Translation of Bharatiya Jyotish Sastra*. R. V. Vaidya, trans., Delhi: Manager of Publications, Civil Lines.

Drake, Stillman (1976). "Galileo's first telescopic observations." *Journal for the History of Astronomy*, 7,153.

Pingree, David (1976). "The recovery of early Greek astronomy from India." *Journal for the History of Astronomy*, 7, 109.

Sarma, K. V., trans. (1956). *The Goladīpikā* by *Parameśvara*. Madras: The Adyar Library and Research Center.

Shukla, K. S. (1957). *The Sūrya Siddhanta with the Commentary of Paramesvara*. Lucknow: Dept. of Mathematics and Astronomy, Lucknow University.

Swerdlow, Noel Mark (1968). *Ptolemy's Theory of the Distances and Sizes of the Planets: A Study of the Scientific Foundations of Medieval Cosmology*. Yale Univ. Ph.D. thesis.

Van Helden, Albert, (1976). *The importance of the transit of Mercury of 1631. Journal for the History of Astronomy*, 7, 1.

Appendix: Statistical Evaluation

In this appendix a simple probabilistic model is used to evaluate whether or not the correlation between modem and *Sūrya-siddhānta* diameters shown in Table 3 could have arisen by chance. First, randomly choose 5 numbers between 0 and B, where B is some fixed positive number. Call these numbers $X_1, \ldots, X_5$, and let them represent the diameters of Mercury, Venus, Mars, Jupiter, and Saturn, as calculated from data in the *Sūrya-siddhānta*. Then randomly choose 5 numbers $Y_1, \ldots, Y_5$ between 0 and B to represent the modern values for these diameters. What is the probability that the Xs will agree with the Ys as well as do the *Sūrya-siddhānta* and modem diameters listed in Table 3?

For each (X,Y), let $\mathrm{P} = 1 - \min(X/Y, Y/X)$. P is a measure of how close X is to Y, and $P = 0$ if

$X = Y$. It is easy to see that if X and Y are chosen independently in $(0,B)$ with a uniform distribution, then P is distributed uniformly on $(0,1)$. (It does not matter what value we choose for B.)

Let S be the sum of the Ps for the 5 pairs (X,Y). If we compute S using the 5 pairs of diameters from Table 3, we get $S = 1.121$. What is the probability that S will be no greater than this for the 5 randomly chosen (X,Y) pairs?

It is easy to compute an upper bound on the probability that $S < Y$, where S

is the sum of *n* independent random variables distributed uniformly on (0,1). This upper bound is $y^n/n!$. Using

S = 1.121 and *n* = 5, we get .0147 for this upper bound. Therefore, the actual pairs of diameters in Table 3 exhibit a significant deviation from chance expectation.

Note that in this probability estimate, the *Sūrya-siddhānta* diameters of Jupiter and Venus have not been doubled. Thus the probability estimate of .0147 is for the unedited *Sūrya-siddhānta* diameters. If we do double the diameters of Venus and Jupiter (taking them to be radii), then the probability estimate becomes 7.7×10^{-6}.

CHAPTER 7

Anomalous Textual Artifacts in Archeo-astronomy

Originally published in:
Revisiting Indus-Sarasvati Age and Ancient India
Edited by Bhu Dev Sharma and Nabarun Gose
Atlanta: World Association for Vedic Studies (1998), pp. 317–334

Based on a conference presentation at:
"Revisiting Indus-Sarasvati Age and Ancient India"
World Association of Vedic Studies (WAVES)
Atlanta, GA / October 1996

Abstract

Ancient artifacts can survive within written texts, as well as within the strata of the earth. In this paper, two examples of anomalous textual artifacts are discussed. The first is a verse from the medieval Indian astronomy text *Sūrya-siddhānta* which seems to encode accurate values for the diameters of the planets. The second is a system of cosmic geography from the *Bhāgavata Purāṇa* which contains what seems to be a realistic map of planetary orbits in the solar system. We discuss these examples in relation to the controversial claim that there existed a pre-modern civilization with advanced astronomical knowledge.

Introduction

Reconstructions of ancient history are necessarily based on fragmentary artifacts and documents that have survived wars, social upheavals, and processes of gradual attrition. In some cases, extensive technological developments are completely lost to recorded history and then revealed by a chance discovery. For example, in 1900 an astronomical computer was uncovered from a shipwreck of the 1st century B.C. near the Greek island of Antikythera (de Solla Price, 1962). This machine used an ingenious system of gears to exhibit positions of the sun, moon, and probably the planets on a series of dials. Nothing like it is described in surviving ancient literature, but a machine of such sophistication implies the existence of a well-developed technological tradition.

In India, in 1958, a realistic cast head of what looks like a Vedic warrior was rescued from being melted down (Hicks and Anderson, 1990). It has been dated to 3700 ± 800 B.C. by MASCA corrected carbon-14 dating. At present, it seems to be the only surviving sculpture of its kind. In this case, the find corroborates popular traditions about ancient Vedic civilization, but it stands out as an anomaly in established reconstructions of ancient history.

Information in old texts may also sometimes provide a unique glimpse into a forgotten past. One type of textual artifact consists of knowledge that seems too advanced for the historical period of the text. In cases where comparable knowledge was acquired in modern times through extensive scientific efforts, it can be argued that the knowledge may be a remnant from an earlier, advanced civilization that is lost to historical memory. In this paper, we will discuss two examples of textual artifacts that suggest the existence of advanced astronomical knowledge in the distant past.

Planetary Diameters

We begin with a medieval Indian astronomical text called the *Sūrya-siddhānta* (Burgess, 1989). In chapter 7 of this text, the 13th verse gives the following rule for calculating the apparent diameters of the planets Mars, Saturn, Mercury, Jupiter, and Venus: "The diameters upon the moon's orbit of Mars, Saturn, Mercury, and Jupiter, are declared to be thirty, increased successively by half the half; that of Venus is sixty."

The meaning is as follows: The diameters are measured in a unit of distance called the *yojana*, which in the *Sūrya-siddhānta* is about five miles. The phrase "upon the moon's orbit" means that the planets look from our vantage point as though they were globes of the indicated diameters situated at the distance of the moon. (Our vantage point is ideally the center of the earth.) Half the half of 30 is 7.5. Thus the verse says that the diameters "upon the moon's orbit" of the indicated planets are given by 30, 37.5, 45, 52.5, and 60 *yojanas*, respectively.

The next verse uses this information to compute the angular diameters of the planets. This computation takes into account the variable distance of the planets from the earth, but for the purposes of this paper it is enough to consider the angular diameters at mean planetary distances. The diameters upon the moon's orbit were given for the planets at these mean distances from the earth. The *Sūrya-siddhānta* says that there are 15 *yojanas* per minute of arc at the distance of the moon (giving 324,000 *yojanas* as the circumference of the moon's orbit). Thus the mean angular diameters of the planets can be computed by dividing the diameters upon the moon's orbit by 15. Table 2 gives the results of this computation and lists other estimates of planetary angular diameters for comparison.

TABLE 1

Planet	*Sūrya-siddhānta*	Ptolemy	Tycho Brahe	Modern Minimum	Modern Maximum
Mars	2.0	1.57	1.67	.067	.450
Saturn	2.5	1.74	1.83	.250	.350
Mercury	3.0	2.09	2.17	.067	.200
Jupiter	3.5	2.61	2.75	.333	.817
Venus	4.0	3.13	3.25	.150	1.233

Angular diameters of planets in minutes of arc. The modern angular diameters are for the greatest and least distances of the planets from the earth. Those of Claudius Ptolemy are from his book *Planetary Hypotheses*.

The figures of Ptolemy and Brahe were obtained by naked-eye astronomy. They are roughly comparable with the *Sūrya-siddhānta* figures and are clearly much larger than the angular diameters measured in more recent times by means of telescopes (Burgess, 1989). It is well known that a small, distant light source looks larger to the naked eye than it really is. This phenomenon makes it likely that angular diameters of planets would inevitably have been overestimated by astronomers before the age of the telescope. However, there is evidence indicating that the *Sūrya-siddhānta* angular diameters were not simply based on naked-eye observation. To see this, we must consider the orbits of the planets.

Orbital Dimensions in the *Sūrya-siddhānta*

Verses 12.85–90 of the *Sūrya-siddhānta* give the circumferences of the planetary orbits in *yojanas*, and these figures are reproduced in Table 2. The orbits are represented as simple circles centered on the earth, and their circumferences are proportional to the mean orbital periods of the planets. For Mercury and Venus, the mean planetary position is the same as the position of the sun, and thus the orbital circumferences in the table are the same for Mercury, Venus, and the Sun. For Mars, Jupiter, and Saturn, the mean position corresponds to the average motion of the planet in its heliocentric orbit.

Verse 1.59 of the *Sūrya-siddhānta* gives the diameter of the earth as 1,600 *yojanas*. Several scholars have argued that the *yojana* in the *Sūrya-siddhānta* is about 5 miles, thereby bringing the earth's diameter to the realistic value of 5×1,600 = 8,000 miles. Examples are Sarma (1956), Dikshit (1969), and Burgess (1989, p. 44). Using modern data for the mean diameter of the earth, the exact value should be 4.948 miles/*yojana*. We will see later on that it is worthwhile to consider this exact value, rather than the rough estimate of 5 miles.

Different standards were adopted for the *yojana* by different medieval Indian astronomers. This was noted by the astronomer Parameśvara (1380–1450 A.D.), who said that "What is given by Āryabhaṭa as the measure of the earth and the distances [of the planets from it], etc., is given as more than one and a half times by other [astronomers]; this is due to the difference in the measure of the *yojana* [adopted by them]" (Sarma, 1956). In fact, using Āryabhaṭa's earth diameter of 1,051 *yojanas*, we get 7.54 miles/*yojana* (Clark, 1930, p. 15).

Verse 4.1 of the *Sūrya-siddhānta* gives the diameters of the sun and moon as 6,500 and 480 *yojanas*, respectively. Given 4.948 miles per *yojana*, the resulting lunar diameter of 2,375 miles is about 10% higher than the modern value. The corresponding earth-moon distance of about 255,000 miles is high by 7%. However, the sun's diameter comes to 32,164 miles, which is far too small.

It is easy to see why the diameter of the moon should be reasonably accurate. The dimensions of the moon and its orbit were well known in ancient times. For example, the lunar diameter given by Ptolemy in his *Planetary Hypotheses* falls within about 7% of the modern value, if we convert his earth-diameters into miles using the modern diameter of the earth (Swerdlow, 1968).

TABLE 2

Planet	*Sūrya-siddhānta* orbital circumference (*yojanas*)	Modern orbital period (days)
Moon	324,000	27.32166
Mercury	4,331,500	365.256
Venus	4,331,500	365.256
Sun	4,331,500	365.256
Mars	8,146,909	686.98

Planet	*Sūrya-siddhānta* orbital circumference (*yojanas*)	Modern orbital period (days)
Jupiter	51,375,764	4,332.587
Saturn	127,668,255	10,759.2

Geocentric orbital circumferences, as given in texts 12.85–90 of the *Sūrya-siddhānta*. The circumferences are nearly proportional to the geocentric planetary periods listed in the second column.

It is also easy to see why the diameter for the sun is too small. Ancient Greek astronomers tended to greatly underestimate the earth-sun distance, and this also happened in the *Sūrya-siddhānta*. The angular diameter of the sun is easily seen to be about the same as that of the moon – about ½ degree. This angular diameter, combined with a small earth-sun distance, leads inevitably to a small estimate for the diameter of the sun. Ptolemy's solar diameter figure is similar to the *Sūrya-siddhānta's*.

Computing Planetary Diameters

What about the planets? Ptolemy listed wildly inaccurate diameters for Mercury, Venus, Mars, Jupiter, and Saturn in his *Planetary Hypotheses*. To see what the *Sūrya-siddhānta* says about the diameters of these planets, we should scale up the diameters on the moon's orbit by multiplying them by the orbital circumference of the planet, divided by the orbital circumference of the moon. This is done in Table 4. (These results were reported in Thompson (1997), assuming 5 miles per *yojana*, rather than the 4.948 used here.)

TABLE 3

Planet	Modern Diameter	*Surya-Siddhanta* Diameter	% Error
Mercury	3031	2977	–2
Venus	7520	3969	–47
Mars	4218	3773	–11
Jupiter	88729	41195	–54
Saturn	74565	73120	–2

Planetary diameters in miles, computed using the *Sūrya-siddhānta* orbital radii from Table 2 and the rule of verse 7.13. The error percentages compare the *Sūrya-siddhānta* diameters with the corresponding modern planetary diameters.

Note that even though the angular diameters are too large, and the orbital radii are too small, the calculated diameters are close to modern values for Mercury,

Mars, and Saturn. For Venus and Jupiter, they are too small by about 50%, and it appears that they may represent radii. As we will see below, in this case they differ from modern values by 6% and -7%.

One might argue that this balancing is due to pure chance. However, the balancing works for five distinct cases. It is therefore worthwhile to estimate just how probable it is.

This probability can be evaluated by randomly choosing 5 pairs of numbers representing modern and ancient planetary diameters. For each pair (x, y), the number $P = 1 - \min(x/y, y/x)$ is a measure of how close x is to y. The sum of the P's for all 5 pairs is a measure of how well they agree. This sum can be computed for the 5 pairs in Table 3, and we can calculate how probable it is that randomly chosen pairs will give a lower sum and a better match. This probability comes to .0175, and therefore there is at most 1 chance in 57 that we would get the results in Table 3 by chance.

If the observed correlation did not happen by chance, then perhaps it happened by design. One hypothesis is that at some time in the past, ancient astronomers possessed realistic values for the diameters of the planets. They might have acquired this knowledge during a forgotten period in which astronomy reached a high level of sophistication. Later on, much of this knowledge was lost, but fragmentary remnants were preserved and eventually incorporated into texts such as the *Sūrya-siddhānta*. In particular, the real diameters of the planets were later combined with erroneous orbital circumferences to compute the diameters "upon the moon" given in verse 7.13. These figures were then accepted because they gave realistic values for the angular diameters of the planets as seen by the naked eye.

This hypothesis is supported by the fact that the *Sūrya-siddhānta* diameters of Jupiter and Venus in Table 2 are almost exactly half of the corresponding modern diameters. If we multiply these *Sūrya-siddhānta* diameters by 2, we get 82,390 miles for Jupiter and 7,939 miles for Venus. These figures differ from the corresponding modern values by 6% and -7%. Given this correction, the root-mean-square error for all five planets comes to 6.6%.

One can argue that the *Sūrya-siddhānta* diameters for Jupiter and Venus were actually the radii for these planets, and somehow they were accepted as diameters by mistake. Or radii might have been deliberately used instead of diameters in order to allow for the simple rule of $30+7.5i$ used in verse 7.13. This is consistent with the fact that such verses were intended as memory aids and brevity was considered to be a virtue.

In the probability estimate discussed above, the *Sūrya-siddhānta* diameters of Jupiter and Venus were not interpreted as radii. Thus the probability estimate of .0175 is for the unedited *Sūrya-siddhānta* diameters. If we do double the diameters of Venus and Jupiter (taking them to be radii), then the probability estimate becomes 1.3×10^{-5}, or about 1 chance in 75,000.

Alternative Explanations

One possible explanation is that verse 7.13 may have been written recently, using modern planetary data, and falsely interpolated into the text. But this is ruled out by the fact that there is a manuscript of the *Sūrya-siddhānta* that scholars date to the year A.D. 1431 (Shukla, 1957). This manuscript includes a commentary by Parameśvara, who died in A.D. 1450, and thus it definitely dates back to the 15th century. Verse 7.13 is present in this manuscript, and it agrees with the Burgess translation quoted above. The commentary explains the verse point by point, and thus it confirms that the verse was present in the manuscript in the same form in which it appears today.

In 15th century Europe, the prevailing ideas concerning the sizes of the planets came from medieval Islamic astronomers who were following the teachings of Ptolemy. The first telescopic observations of planets were made by Galileo in 1609–10 (Drake, 1976). As late as 1631, Pierre Gassendi of Paris was shocked when his telescopic observation of a transit of Mercury across the sun revealed that its angular diameter was much smaller than he had believed possible (Van Helden, 1976). It is clear that the information on planetary diameters in the *Sūrya-siddhānta* antedates the development of modern knowledge of these diameters.

It is also clear that Hellenistic astronomers did not have accurate diameters for the planets. Ptolemy computed planetary diameters from his angular diameters, and his estimates of planetary distances, and these were reproduced without significant change by European and Islamic astronomers for centuries (Swerdlow, 1968). However, his figures disagree strongly both with modern data and with the diameters computed from *Sūrya-siddhānta* in Table 3.

If we hypothesize that verse 7.13 incorporates knowledge of the actual diameters of the planets, then one natural question is this: If one started with the modern diameters of the planets and the *Sūrya-siddhānta* orbital circumferences, could one calculate backwards and arrive at the rule given in this verse? It turns out that the answer is yes. For each planet, multiply its modern diameter, converted to *yojanas*, by the ratio between the orbital circumferences of the moon and the given planet, as listed in Table 2. Here we use the radius in place of the diameter for Jupiter and Venus. The resulting values can be approximated by an arithmetic progression of the form $ai + b$ either by trial and error or by using an optimization method such as least squares. If we divide by 15 to get minutes of arc and then round off, we get the angular diameters given by *Sūrya-siddhānta*.

Planetary Orbits

The *Sūrya-siddhānta* makes the planetary orbits too small, and this was also done by ancient Greek astronomers. However, it turns out that in the *Bhāgavata Purāṇa*, the orbits of the planets are implicitly represented with good accuracy.

The *Bhāgavata Purāṇa* (*Bhāgavatam*) is a Sanskrit text sacred to the Vaiṣṇavas or worshipers of Viṣṇu. Modern scholars generally date it somewhere between the 5th and the 10th centuries A.D., but Indian tradition assigns it to about 3000 B.C.

The *Bhāgavatam* contains a section (the Fifth Skanda) which deals with cosmology. This describes the earth as a disk, called Bhū-maṇḍala or earth mandala, which is 500 million *yojanas* in diameter. This disk is divided into a series of concentric ring-shaped oceans and islands. The islands, called *dvīpas*, are further subdivided by mountains, rivers, and other geographical features. The radii of the *dvīpas*, oceans, and ring mountains are given in the text in *yojanas*.

TABLE 4

N	Radius	Thickness	Feature
1	50	50	Jambūdvīpa
2	150	100	Salt Ocean
3	350	200	Plakṣadvīpa
4	550	200	Sugarcane Ocean
5	950	400	Śālmalidvīpa
6	1350	400	Liquor Ocean
7	2150	800	Kuśadvīpa
8	2950	800	Ghee Ocean
9	4550	1600	Krauñcadvīpa
10	6150	1600	Milk Ocean
11	9350	3200	Śākadvīpa
12	12550	3200	Yogurt Ocean
13	15750	3200	Manasottara Mt.

N	Radius	Thickness	Feature
14	18950	3200	Puṣkaradvīpa
15	25350	6400	Sweet Water Ocean
16	41100	15750	Inhabited Land
17	125000	83900	Golden Land
18	250000	125000	Aloka-varṣa

The radii in thousands of *yojanas* of the circular features of Bhū-maṇḍala, as given in the *Bhāgavata Purāṇa*.

At first glance, Bhū-maṇḍala seems to be a poetic but utterly mythological description of the earth as a flat disk. However, the sizes of the *dvīpas* have a curious feature. They are far too small for the universe of stars and galaxies and far too large for the earth globe. But, given traditional values for the length of the *yojana*, they look right for the solar system.

The *Bhāgavatam* describes the path of the sun over Bhū-maṇḍala in detail, and from this one can deduce that Bhū-maṇḍala corresponds roughly with the plane of the ecliptic in modern astronomy. It is therefore natural to ask how the orbits of the planets compare with the geographical features of Bhū-maṇḍala.

Since the earth as we know it corresponds with the center of the Bhū-maṇḍala disk, we need to look at planetary orbits from a geocentric point of view. For each planet, we did this by taking the planet's heliocentric position vector and subtracting the earth's heliocentric position vector. By using an up-to-date ephemeris program, one can thus obtain the geocentric orbits of the planets as given by modern astronomy.

To compare the geocentric orbits of the planets with the *dvīpas* of Bhū-maṇḍala, we need to convert the *dvīpa* radii from *yojanas* to miles or kilometers. We have already observed that there are different standards for the length of the *yojana*, with 4.948 miles bringing the earth to its correct diameter for *Sūrya-siddhānta*.

Hiuen Thsang, a Buddhist pilgrim who visited India in the 7th century A.D., said that a *yojana* consists of 40 *li* according to tradition, but the measure in customary use was equal to 30 *li*, and the measure given in sacred texts was only 16 *li* (Burgess, 1989, p. 43). Joseph Needham (1962, p. 52) pointed out that the *li* has taken many values during China's history, but in the Tang dynasty (when Hiuen Thsang lived), there was a long *li* of 532 meters and a short *li* of 442 meters. This gives about 4.4 to 5.3 miles for the 16-*li yojana* and 8.2 to 9.9 miles for the 30-*li yojana*. These figures roughly agree with the *yojana* of the

Sūrya-siddhānta and with the popular valuation of 8 to 9 miles per *yojana* given by Cunningham (1990, p. 486). Of course, the *yojana* of 40 *li* is substantially larger than these values.

If we use a value of about 8 miles per *yojana*, we find that the geocentric orbits of Mercury, Venus, Mars, Jupiter, and Saturn, projected to the plane of the ecliptic, are in striking agreement with the features of Bhū-maṇḍala. We observe that the apogees and perigees of these orbits (their points of greatest and smallest distance from the earth) seem to line up with the boundaries of *dvīpas* and oceans, as listed in Table 4.

A geocentric orbit looks like a spirograph tracing that loops around the earth (see Figures 2–5). There are points of closest approach, where the curve swings in toward the center, levels off, and then moves out. These points of closest approach range from a minimum (the perigee) to a maximum. Likewise, the points of greatest distance, where the orbit turns in after moving away, range from a minimum to a maximum (the apogee). These four minima and maxima can be called the turning points of the orbit, and they help describe what the orbit looks like when plotted on a sheet of paper.

We can define a measure of "goodness of fit," showing how well the orbits line up with the *dvīpa* radii given in the *Bhāgavatam*. For each planet, find the shortest distance between a *dvīpa* radius and an orbital turning point. Take the root mean square of these distances for Mercury, Venus, Mars, Jupiter, and Saturn. The reciprocal of this is large if the orbits line up well with *dvīpas*, and it is small if they do not line up well.

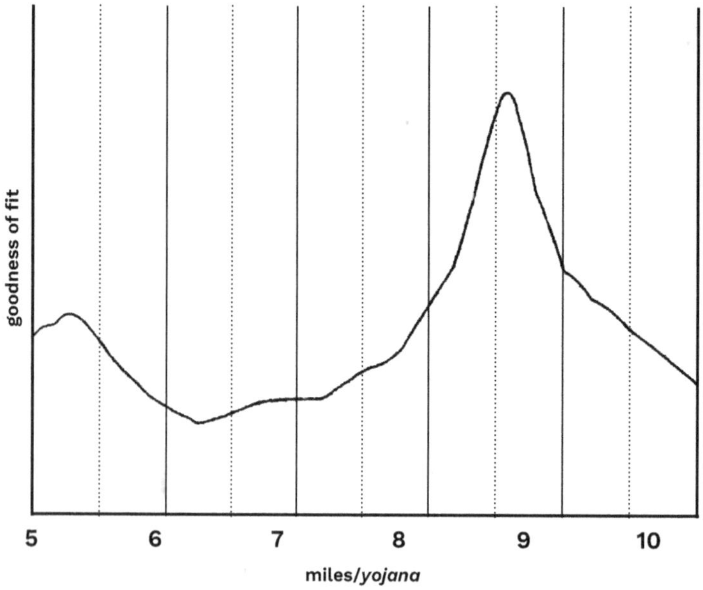

Figure 1. Plot of goodness of fit as a function of miles/*yojana* in the range 5 to 10. The peak is at 8.575 miles/*yojana*.

In Figure 1, this reciprocal root mean square is plotted as a function of the number of miles per *yojana*, which is allowed to range from 5 to 10. We can see that the curve has a pronounced peak at 8.575 miles/*yojana*. Thus, if we search for a good fit between orbits and *dvīpas*, we automatically arrive at a value for the number of miles/*yojana* that agrees with customary usage and with the *yojana* of 30 *li*.

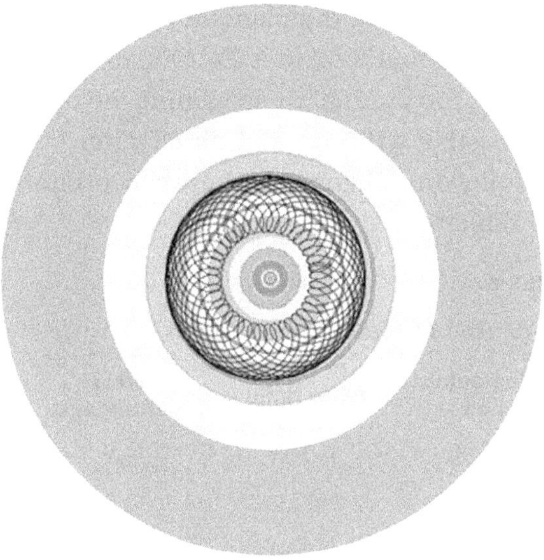

Figure 2. Plot of the geocentric orbit of Mercury (spirograph) and the geocentric orbit of the sun (dark circle) superimposed on Bhū-maṇḍala. The epoch is 3102 B.C., and the plot assumes 8.575 miles/*yojana*.

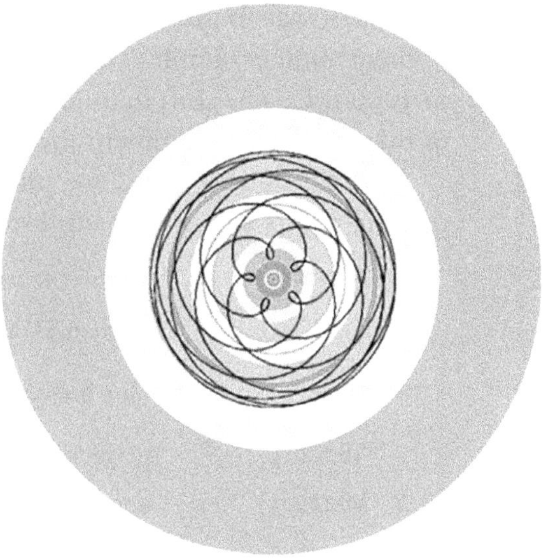

Figure 3. Plot of the geocentric orbit of Venus (spirograph) and the geocentric orbit of the sun (dark circle) superimposed on Bhū-maṇḍala.

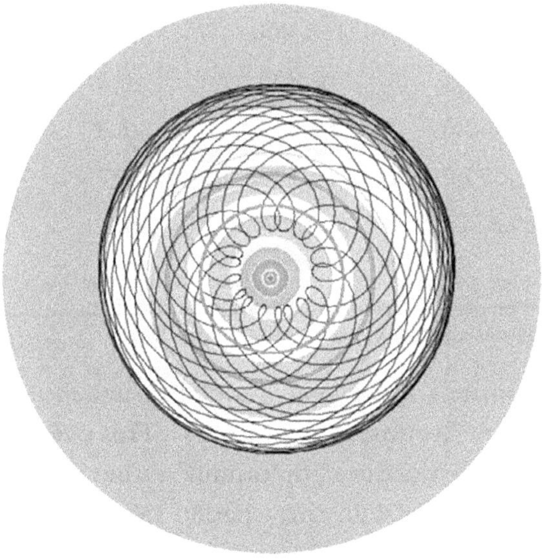

Figure 4. Plot of the geocentric orbit of Mars superimposed on Bhū-maṇḍala.

Figure 5. Plot of the geocentric orbit of Jupiter (inner) and Saturn (outer) superimposed on Bhū-maṇḍala.

This calculation makes use of orbital calculations for the epoch of 3102 B.C., the traditional starting date of Kali-yuga. This date was chosen since it marks the traditional time period of the civilization described in the *Bhāgavatam*. In general, it was found that, due to gradual shifting of the planetary orbits, the optimal number of miles/*yojana* changes from about 8.52 in A.D. 2000 to 8.605 in 4000 B.C. The shift in miles per *yojana* is too gradual to determine the date of the orbit/*dvīpa* alignment, but it is consistent with the traditional date.

At 8.575 miles/*yojana*, we find the following alignments between orbital turning points and *dvīpa* radii listed in Table 5.

TABLE 5

Correlation Between Dvipa Radii And Orbital Turning Points

N	Planet	turning point	turning point radius	*dvīpa* radius	error percentage
1	Mercury	perigee	5,976.0	6,150	2.9
2	Mercury	apogee	15,701.1	15,750	0.3
3	Venus	perigee	2,851.0	2,950	3.5
4	Venus	apogee	18,813.0	18,950	0.7
5	Mars	perigee	4,090.0	4,550	11.2
6	Mars	apogee-	25,736.5	25,350	-1.5
7	Jupiter	perigee	43,422.8	41,100	-5.3
8	Saturn	apogee	121,599.6	125,000	2.8
9	Sun	mean	10,840.4	10,800	-0.4
10	Ceres	perigee	16,312.8	15,750	-3.4
11	Ceres	apogee	42,683.2	41,100	-3.7
12	Uranus	apogee	229,811.0	250,000	8.8

The *dvīpa* radii arc from Table 4. Error percentage gives the error in the *dvīpa* radius relative to the corresponding orbital turning point. The orbital turning points are calculated for the beginning of Kali-yuga, using the ephemeris programs of Duffett-Smith (1985).

We should note that calculation for the optimal miles/*yojana* took into account the radii of Table 4, except for Mānasottara Mountain (feature 13). This oversight was brought about by the fact that this feature is a "mountain" rather than a boundary of a ring-shaped region. It is one of two ring-shaped mountains in the Bhū-maṇḍala system (the other being Lolāloka Mountain marking the outer boundary of the Golden Land, feature 17). However, it turns out that this mountain closely matches the apogee of Mercury. Thus the perigees and apogees of Mercury and Venus are all aligned closely with circular features of Bhū-maṇḍala.

In the case of Mars, the minimum of the envelope of outer turning points lines up with the outer boundary of the Sweet Water Ocean (feature 15). This is called apogee – to distinguish it from the apogee proper, which is the maximum of the outer turning points. The perigee of Mars lines up with the outer boundary of Krauñcadvīpa (feature 9), but the error is greater than that of the other correlations, and Mars partially penetrates the interior of Krauñcadvīpa. (Perhaps coincidentally, the *Bhāgavata* 5.20.19 states that Krauñcadvīpa was attacked by Kārttikeya, who is the regent of Mars.)

The perigee of Jupiter and the apogee of Saturn line up with the inner and outer boundaries of the region known as Golden Land (features 16 and 17). The outer boundary – called Lokāloka Mountain – is said to divide the regions lit by the sun from those not lit by the sun (*Bhāgavata* 5.20.34). Saturn, the outermost of the five traditional planets, orbits just within this boundary.

The sun, of course, is one of the most important elements of *Bhāgavata* cosmology. The geocentric orbit of the sun is nearly circular and falls close to the midpoint of the Yogurt Ocean (feature 12). This midpoint comes to 10,950 = (9,350 + 12,550)/2 thousand *yojanas*.

When these correlations emerged, it was natural to ask about the other planets in the solar system. Neptune and Pluto lie outside the outer boundary of Bhū-maṇḍala, but the apogee of Uranus is within this boundary (feature 18) and close to it. Even the asteroid Ceres lines up with *dvīpa* boundaries in Bhū-maṇḍala. Ceres is the largest asteroid, and it marks the orbit of the famous missing planet between Mars and Jupiter. Since the orbital elements of Uranus and Ceres are not as well known as those of the other planets, the calculations for Uranus and Ceres were made for the epoch of A.D. 2000, rather than the third millennium B.C. (using, as always, 8.575 miles/*yojana*).

We made an effort to test these correlations statistically. The strategy is to focus attention on the correlations that emerged from the original optimization program used to find miles/*yojana*. For the purpose of statistical testing, other correlations that were noticed later were ignored.

The statistical test was performed by generating random sets of concentric circles and comparing them with the geocentric orbits of the planets. To keep these similar to the actual Bhū-maṇḍala system, the thicknesses from Table 4 were multiplied by pseudo-random numbers in the range of 0.25 to 1.75, and pseudo-radii were computed by adding these thicknesses. The optimal miles/*yojana* figure within the range 5 to 9 was then computed for the five planets

Mercury through Saturn, and the reciprocal root mean square measure of correlation was also computed. This process was repeated 10,000 times.

It was found that in 9 cases out of 10, the level of correlation was less than that obtained for the actual Bhū-maṇḍala system. Thus the original pattern of alignments found by the optimization program is statistically significant. The additional alignments observed later can be regarded as icing on the cake.

The Length of the *Yojana*

The value of 8.575 miles/*yojana* was calculated by fitting orbits to *dvīpas*, but it turns out to be closely related to the dimensions of the earth. There are 68.70843 miles per degree of latitude at the equator. At 8.575 miles/*yojana*, we have 8.013 *yojanas* per degree. Since this is very close to 8, it is possible that the *yojana* was intended to be 1/8 of a degree of latitude at the equator, or 8.589 miles. This would be consistent with the fact that units of distance have often been defined in terms of latitude, both in modern and in ancient times. This is true of the meter (which is close to 1/10,000,000 of the distance from the equator to the pole on the Paris meridian) and the Greek stadium (which was 1/600 of a degree of latitude).

The *yojana* is often stated to be four *krośas* of 2000 or 1000 *dhanus* (bow-lengths). A *dhanu* is 4 *hastas*, where a *hasta* of 24 *aṅgulas* (fingers) corresponds to a Western cubit (Burgess, 1989, p. 43). The *yojana* thus comes to 32,000 or 16,000 *hastas*. Let us refer to these units as a long and short *yojana*, respectively.

If a long *yojana* is defined as 1/8 of a degree of latitude, then a *hasta* comes to 431.936 millimeters. I would like to suggest that the long *yojana* may correspond to the 30-*li yojana* mentioned by Hiuen Thsang, and the short *yojana* may correspond to his 16-*li yojana*. If so, the *hasta* of the short *yojana* must be 32/30 times the *hasta* of the long *yojana*. This comes to 460.732 millimeters.

This is within 0.2 percent of the "geographic cubit" which was used extensively in ancient Near Eastern and Western civilizations for geographical measurements. Livio Stecchini, a scholar specializing in the history of measures, has explained in detail the use of the geographic cubit in ancient Egypt, and he defines it as 461.6935 millimeters (Stecchini, 1971, p. 351). He also points out that the Greek stadium was 400 "Greek cubits" of 462.4147 millimeters, and was considered to be 1/600 of a degree of latitude.

Stecchini argues that the difference between these two cubits is due to the variation of the length of a degree of latitude at different latitudes. The Greek cubit corresponds to the latitude of Mycenae, and the latitude for the geographic cubit lies in Egypt. For comparison, we have defined the *hasta* of 460.732 millimeters in terms of the degree of latitude at the equator. This agrees well with our study of planetary orbits.

The connection between the *yojana* and the Greek and geographic cubits is corroborated by Strabo, who describes the experiences of Megasthenes, a Greek ambassador to India in the period following Alexander the Great. Strabo cites Megasthenes as saying that along the royal road to the Indian capital of Palibothra (thought to be modern Patna), there were pillars set up every 10 *stadia*. The British scholar Alexander Cunningham argues that the pillars marked an interval of one *krośa* (Cunningham, 1990, p. 484). Since the *stadium* was defined as 400 Greek cubits, there must be 4,000 cubits per *krośa* and 16,000 per *yojana*. This agrees nicely with our conclusion that the short *yojana* should consist of 16,000 of the geographic or Greek cubits.

According to this interpretation, the pilgrim Hiuen Thsang learned the proportions of three sizes of *yojanas* from his Indian hosts, but he expressed their absolute lengths only approximately in his native *li*. (Of these, the first two are our short and long *yojanas*, and the third remains unidentified.) This might be expected of a pilgrim, whose concern with units of distance was to provide travel directions for his fellow pilgrims.

In contrast, the close agreement between 10 *stadia* and the *krośa* of the short *yojana* suggests a long-standing connection between India and the West. Note that since the *stadium* was 1/600 of a degree of latitude, the *krośa* of the short *yojana* comes to 1/60 of a degree, or 1 minute. The short *yojana* itself is 1/15 of a degree.

As we have already noted, the *hasta* of the short *yojana* is 32/30 of the 431.936 millimeter hasta of the long *yojana*. This number of millimeters, in turn, is very close to 4 times 108, a number that shows up repeatedly in ancient Indian literature and in the ancient world in general.

Here a key observation is that the meter is very close to 1/10,000,000 of the length of the meridian from the equator to the north pole. (It was fixed at this value after the French Revolution, but later remeasurements led to a slight revision in the length of the meridian in meters.) Thus, the *hasta* of the long *yojana* is close to 432 ten billionths of the length of this meridian, or 108 ten billionths of the circumference of the earth through the poles.

In a perfect sphere, the circumference through the poles is 360 times the degree of latitude at the equator. But in the oblate spheroid of the earth, this ratio is slightly higher. We can approximate it by taking a degree of latitude at the equator to be 8 long *yojanas* of 32,000 *hastas* of 108 ten billionths of the circumference through the poles. This gives us a value of 361.690 = 10,000,000/(256 × 108), where 256 is 2 to the 8th power. In fact, this simple estimate is low by only 0.04%.

Another expression for the equatorial bulge can be derived by considering the *yojana* of 4.948 miles that figured in the study of planetary diameters in the *Sūrya-siddhānta*. We pointed out above that the figure of 4.948 miles would prove to be important. Recall that this figure was obtained by dividing the exact mean diameter of the earth in miles by 1,600, the number of *yojanas* in the diameter of the earth according to the *Sūrya-siddhānta*.

Now, 4.948 differs substantially from the short *yojana* of 4.58 *yojanas*. However, 4.948/4.58 is within 0.03% of 1.08. Whether this is a coincidence or not, it is a fact that 1,600 × 1.08 short *yojanas* is very close to the mean diameter of the earth. The short *yojana* is 8/15 of the long *yojana*, which we just expressed as a fraction of the circumference of the earth through the poles. By multiplying these numbers together, we can estimate the ratio between the polar circumference and the earth's mean diameter to be 150,000,000/(4,096 × 108 × 108), where 4,096 is 2 to the 12th power. This estimate comes to 3.13967, a number that is close to π, but slightly lower, due to the earth's equatorial bulge.

We can also compute this ratio using calculus and the flattening factor of $f = 1/298.257$ from a modern spherical astronomy text (Green, 1985, p. 100). The result of this calculation is 3.13984, and it is 0.006% higher than the estimate obtained from the study of the *yojana*. It is not likely that one could come so close to this number simply by chance. It appears that the *yojana* in its different forms reflects real knowledge of the dimensions of the earth.

In view of these findings, we should briefly mention the role of 108 in India and other parts of the world. The Egyptologist Jane Sellers (1992) has documented the repeated appearance of 108, 432, 360, and related numbers in many different ancient and medieval traditions. These numbers also show up over and over again in ancient Indian texts, such as the *Ṛg Veda* (Kak, 1987, 1993).

Sellers interprets these numbers as evidence for ancient knowledge of precession of the equinoxes. In Indian texts there may also be a connection with precession, but it is clear that these numbers play a variety of astronomical roles. For example, in the *Bhāgavata Purāṇa*, the period of Jupiter is given very simply as 12 × 360 = 4,320 days and the period of Saturn is given as 30 × 360 = 10,800 days (the actual figures being 4332.6 days and 10,759.2 days). The additional role of 108 in the *yojana* and the dimensions of the earth suggests that it may have played an important role in a sophisticated ancient system of mathematics and astronomy.

As a final point, recall that the mean earth-sun distance listed in Table 5 is

10,840.5 thousand *yojanas* and is close to the 10,950 thousand-*yojana* midpoint of feature 12. Here we have another possible example involving 108 and 360. Note that 10,800 = 30 × 360, and 10,950 = 30 × 365.

Conclusions

There is evidence that Bhū-maṇḍala, as described in the *Bhāgavata Purāṇa*, can be interpreted as an accurate map of the solar system showing how the planets move relative to the earth. This map is expressed in terms of distance units based on accurate knowledge of the dimensions of the earth.

There is also evidence that realistic knowledge of the diameters of the planets lies behind text 7.13 of the *Sūrya-siddhānta*. This knowledge is expressed in a unit of measurement that seems to be related to the unit used in the *Bhāgavatam*, and the relationship leads to an accurate expression for the ratio between the circumference of the earth through the poles and the earth's mean diameter.

This evidence of advanced astronomical knowledge suggests the existence of a civilization with well developed sciences of astronomy and geography. But when could such a civilization have flourished? The *Sūrya-siddhānta* and other medieval Indian astronomy texts strongly underestimate the distances of the planets, and this was also done by Greek astronomers from Aristarchus to Ptolemy.

The astronomy of Bhū-maṇḍala in the *Bhāgavatam* was not understood by medieval Indian astronomers. Thus Bhāskarācārya, the 11th-century author of *Siddhānta-śiromaṇi,* said that he could not reconcile the relatively small diameter of the earth globe with the immense size attributed to the earth by the *Purāṇas* (Wilkinson, 1861, pp. 114–115). Likewise, the 15th-century astronomer Parameśvara stated that the Purāṇic account of seven *dvīpas* and oceans is something "given only for religious meditation" and is not acceptable to astronomers (Sarma, 1956, p. 85). Since these astronomers severely underestimated planetary distances, they could not see the correspondence between the "earth maṇḍala" and the solar system.

The known astronomy of the first millennium B.C. and earlier is even more rudimentary than that of the Greeks or medieval Indians (Neugebauer, 1975). Thus, if accurate information about planetary distances and diameters was known in the past, then it appears that this knowledge must date back well before the first millennium B.C. It must be coming down to us in fragmentary form from an earlier high civilization that was followed by a period of collapse and eventual recovery.

The connection between the *yojana* and the geographical cubit may indicate a tradition of knowledge dating back at least as far as the building of the great pyramid in about 2500 B.C. The correlation of planetary orbits with Bhū-maṇḍala gives a *yojana* that is exactly 1/8 of a degree of equatorial latitude at some time in the 4th millennium B.C. (This is due to the slow shifting of the orbits with the passage of time.) These indications of a date between 2500 and 4000 B.C. are consistent with the evidence indicating that astronomical knowledge was crude from the first millennium B.C. up to the medieval period.

In recent times, knowledge of planetary distances and diameters was obtained using the telescope, aided by Newtonian physics. It is not apparent how equivalent knowledge may have been obtained in the distant past. However, further research may yield greater insight into the advanced scientific achievements of ancient civilizations.

References

Burgess, Ebenezer, trans., 1989. *The Sūrya Siddhānta*, Gangooly, P., ed., Delhi: Motilal Banarsidass.

Cunningham, Alexander, 1990. *The Ancient Geography of India,* Delhi: Low Price Publications.

de Solla Price, Derek, March 1962. "Unworldly Mechanics," *Natural History,* 71:8–17

Dikshit, S. B., 1969. *English Translation of Bharatiya Jyotish Sastra,* R. V. Vaidya, trans., Delhi: Manager of Publications, Civil Lines.

Drake, Stillman, 1976. "Galileo's First Telescopic Observations," *Journal for the History of Astronomy*, 7, 153–168.

Duffett-Smith, Peter, 1985. *Astronomy with Your Personal Computer*, Cambridge: Cambridge University Press.

Green, Robin, 1985. *Spherical Astronomy,* Cambridge: Cambridge University Press.

Hicks, Harry and Anderson, Robert, 1990. "Analysis of an Indo-European Vedic Aryan Head – 4th Millenium B.C.," *Jour. of Indo-European Studies,* vol. 18.

Kak, Subash, 1987. "On Astronomy in Ancient India," *Indian Jour. of History of Science,* 22(3): 205–221.

Kak, Subash, 1993. "The Structure of the Ṛgveda," *Indian Jour. of History of Science,* 28(2): 71–79.

Needham, Joseph, 1962. *Science and Civilisation in China*, vol. 4, part I, Cambridge: Cambridge University Press.

Neugebauer, 0., 1975. *A History of Ancient Mathematical Astronomy*, Springer-Verlag.

Sarma, K.V., trans., 1956–1957. *The Goladīpikā by Parameśvara*, Madras: The Adyar Library and Research Center.

Sellers, Jane B., 1992. *The Death of Gods in Ancient Egypt,* London: Penguin.

Shukla, K. S., 1957. *The Sūrya Siddhānta with the commentary of Parameśvara,* Lucknow: Dept. of Mathematics and Astronomy, University of Lucknow.

Stecchini, Livio Catullo, 1971, "Notes on the Relation of Ancient Measures to the Great Pyramid," in Tompkins, Peter, *Secrets of the Great Pyramid,* New York: Harper & Row, 287–382.

Swerdlow, Noel Mark, 1968. *Ptolemy's Theory of the Distances and Sizes of the Planets: A Study of the Scientific Foundations of Medieval Cosmology*, Yale Univ. Ph.D. thesis.

Thompson, Richard, 1997. "Planetary diameters in the *Sūrya-siddhānta,*" *Jour. of Scientific Exploration.*

Van Helden, Albert, 1976. "The Importance of the Transit of Mercury of 1631," *Jour. for the History of Astronomy*, 7: 1–10.

Wilkinson, L., trans., 1861. *Siddhānta-śiromaṇi of Bhāskarācārya,* B. D. Sastrin (Rev.), Calcutta: Baptist Mission Press, reprinted in *Bibliotheca Indica*, New Series No. 1, Hindu Astronomy I.

PART II

PAPERS AND PRESENTATIONS

CHAPTER 8

Points Concerning the Vedic Conception of the Universe

Located in the
Richard L. Thompson Archives
Box 1, Folder 2 (c. 1983)

Notes:

This section offers a forty-one-page document that had been hand-written by Thompson on legal-sized, yellow lined paper, also known as a legal pad. A facsimile reprint of the original document is presented on the left-hand even-numbered pages, with the ride-hand side presenting typewritten text mirroring the facsimile document.

Though undated, evidence suggests it was produced in the early 1980s. The document was initially located in a folder along with copies of thirty typed letters having an average date of 1982. Around this time, digital word processors and other computer programs would become more widely available, and Thompson would utilize them in future work.

This document not only appears to provide a framework for much of Thompson's research on Vedic cosmology, it in good part anticipates more recent scholarship geared toward projects such as the Temple of the Vedic Planetarium constructed in Mayapur, West Bengal, India. It certainly forms a general outline for the contents of his first book on the subject, *Vedic Cosmography and Astronomy,* initially published by the Bhaktivedanta Book Trust in 1990.

m ①

Points concerning the Vedic conception of the universe.

1. We have two main sources of information on this subject:
   (a) the 5th Canto of Srimad Bhagavatam

   (b) the Surya-siddhanta.

2. Under category (a) there are also the accounts of the universe given in the other Puranas, the Mahabharata, and the Ramayan. These are similar to the account given in the 5th Canto of S.B.
   * We should systematically study these accounts in order to see just how similar they are to the 5th Canto. Questions: Do any of these accounts suffer from interpolation or other forms of textual corruption? Also, do any of them refer to the earth as a globe?

3. Under category (b) there are the Jyotisa Vedanta (a short and fragmentary text from the Vedas), the other astronomical "siddhantas", such as the Romaka-siddhanta (which may not be extant), + the works of historical Indian astronomers such as Bhaskaracarya,

Points Concerning the Vedic Conception of the Universe

1. We have two main sources of information on this subject:

   (a) the 5th Canto of *Śrīmad-Bhāgavatam*

   (b) the *Sūrya-siddhānta.*

2. Under category (a) there are also the accounts of the universe given in the other *Purāṇas*, the *Mahābhārata*, and the *Rāmāyaṇa.* These are similar to the account given in the 5th Canto of *S.B.*

   ⋆ We should systematically study these accounts in order to see just how similar they are to the 5th Canto. Questions: Do any of these accounts suffer from interpolation or other forms of textual corruption? Also, do any of them refer to the earth as a globe?

3. Under category (b) there are the *Jyotiṣa Vedānta* (a short and fragmentary text from the *Vedas*), the other astronomical *Siddhāntas*, such as the *Romaka-siddhānta* (which may not be extant), and the works of historical Indian astronomers such as Bhāskarācārya, Brahmagupta, and Āryabhaṭa (who lived in the 1st millennium A.D.).

②

Brahmagupta and Aryabhata.
(who lived in the 1st millenium A.D.)

4. Srila Prabhupada has translated the 5th Canto of S.B., and he has mentioned in the Caitanya-caritamrta that the Surya-siddhanta is authentic. (Srila Prabhupada wrote that Srila Bhaktisiddhanta Sarasvati "wrote" and "compiled" the Surya-siddhanta. However, the Surya-siddhanta is an ancient Sanskrit work, and it appears that Srila Bhaktisiddhanta translated it into Bengali and [perhaps] wrote a commentary. Srila Bhaktisiddhanta's translation of Surya-siddhanta is mentioned in the latest edition of Ebenezer Burgess's translation of this work.) Srila Prabhupada also refered to the Siddhanta-siromani in two of his purports in the C.C. The Siddhanta-siromani is a recent exposition of Indian astronomy, and is very similar to the Surya-siddhanta.

* We should obtain a copy of Bhaktisiddhanta's translation of S-S [=Surya-siddhanta], and also learn whatever we can about the astronomical views of Bhaktisiddhanta and other Vaisnava acaryas. (In this connection we should see A.K. and Sukhamlal Raya Choudury of N.Y.C.)

4. Śrīla Prabhupāda has translated the 5th Canto of *S.B.*, and he has mentioned in the *Caitanya-caritāmṛta* that the *Sūrya-siddhānta* is authentic. (Śrīla Prabhupāda wrote that Śrīla Bhaktisiddhānta Sarasvatī "wrote" and "compiled" the *Sūrya-siddhānta*. However, the *Sūrya-siddhānta* is an ancient Sanskrit work, and it appears that Śrīla Bhaktisiddhānta translated it into Bengali and [perhaps] wrote a commentary. Śrīla Bhaktisiddhānta's translation of *Sūrya-siddhānta* is mentioned in the latest edition of Ebenezer Burgess' translation of this work.) Śrīla Prabhupāda also referred to the *Siddhānta-śiromaṇi* in two of his purports in the *C.C.* The *Siddhānta-śiromaṇi* is a recent exposition of Indian astronomy, and is very similar to the *Sūrya-siddhānta.*

- ⋆ We should obtain a copy of Bhaktisiddhānta's translation of S.S. [= *Sūrya-siddhānta*], and also learn whatever we can about the astronomical views of Bhaktisiddhānta and other Vaiṣṇava ācāryas. (In this connection, we should see A.K. and Sukhamlal Raya Choudury of N.Y.C.)

③

5. We should regard as parampara those works explicitly endorsed by Srila Prabhupada, and we should consider that other Vedic texts may be contaminated by textual corruption. It would seem that we can regard as authentic:
   (a) the 5th Canto of S.B. and other material on astronomy in Srila Prabhupada's purports.
   (b) the *Surya-siddhanta*.

6. I have a "reasonably good" [but not perfect] translation of S.S. by one Bapu Deva Sastri, dating from the 1860's. There is also a blatantly anti-Hindu translation by the Rev. Ebenezer Burgess.

7. The *Surya-siddhanta* + related texts have been traditionally used by the Vaisnava sampradayas for calendar calculations. Also the date of 5000 years for the age of the Bhagavad-gita is based on the S.S. Here is how this date is obtained: The S.S. provides rules for computing the following mathematical function:

F[the kali-yuga date] = positions of the planets at that date.

Here, "kali-yuga date" means the number of years that have elapsed from the beginning of kali yuga to the present.

5. We should regard as *paramparā* those works explicitly endorsed by Śrīla Prabhupāda, and we should consider that other Vedic texts may be contaminated by textual corruption. It would seem that we can regard as authentic:

(a) The 5th Canto of *S.B.* and other material on astronomy in Śrīla Prabhupāda's purports.

(b) the *Sūrya-siddhānta*.

6. I have a "reasonably good" [but not perfect] translation of *S.S.* by one Bapu Deva Sastri, dating from the 1860s. There is also a blatantly anti-Hindu translation by the Rev. Ebenezer Burgess.

7. The *Sūrya-siddhānta* and related texts have been traditionally used by the Vaiṣṇava *sampradāyas* for calendar calculations. Also the date of 5,000 years for the age of the *Bhagavad-gītā* is based on the *S.S.* Here is how this date is obtained: The *S.S.* provides rules for computing the following mathematical function:

F [the Kali-yuga date] = positions of the planets at that date.

Here, "Kali-yuga date" means the number of years that have elapsed from the beginning of Kali-yuga to the present.

④

Now if we observe the present positions of the planets ~~now~~, we can work backwards to find the present kali yuga date:

F^{-1}[present planetary positions] = present kali-yuga date

This calculation yields:

present kali yuga date = 5,102 years + a certain number of days and hours.

[Here "present" refers to a particular date in the 1950's, I believe.]

8. It follows from (7) that we are already using the S.S. when we refer to the Vedic literatures as 5,000 years old.

8.5 Over ⟶

9. The 5th Canto of S.B. givens, roughly, the following picture of the universe:

(a) large scale:

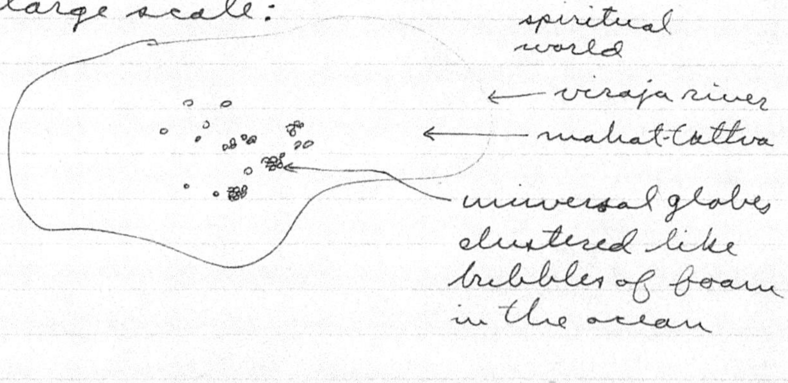

Now if we observe the present positions of the planets, we can work backwards to find the present Kali-yuga date:

F^{-1} [present planetary positions] = present Kali-yuga date

This calculation yields:

> present Kali-yuga date = 5,102 years + a certain number of days and hours.
>
> [Here "present" refers to a particular date in the 1950s, I believe.]

8. It follows from (7) that we are already using the *S.S.* when we refer to the Vedic literatures as 5,000 years old.

8.5. Over

9. The 5th Canto of S.B. gives, roughly, the following picture of the universe:

(a) large scale:

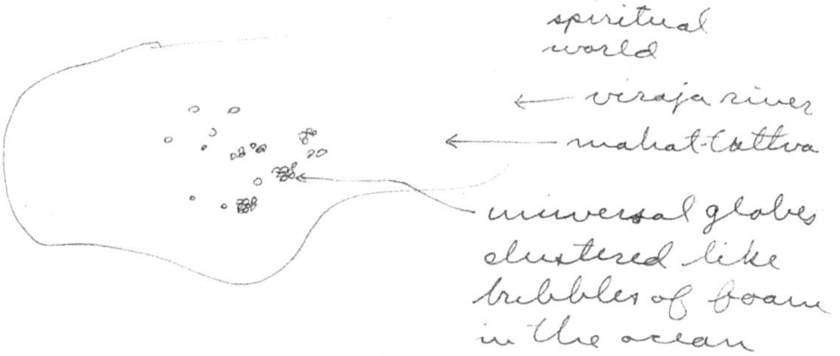

8.5 Srila Prabhupada asked us, in a letter, to make a model of the universe that demonstrates eclipses, the passing of the seasons, and day and night. The 5th Canto provides no basis for the numerical prediction of eclipses, and (as we shall see) it does not even provide an explanation of day and night. However, the Surya-siddhanta does provide ~~...~~ explicit calculations for all of these things. This is another reason for thinking that we should take the Surya-siddhanta seriously.

8.5. Śrīla Prabhupāda asked us, in a letter, to make a model of the universe that demonstrates eclipses, the passing of the seasons, and day and night. The 5th Canto provides no basis for the numerical prediction of eclipses, and (as we shall see) it does not even provide an explanation of day and night. However, the *Sūrya-siddhānta* does provide explicit calculations for all of these things. This is another reason for thinking that we should take the *Sūrya-siddhānta* seriously.

(b) one universal globe or brahmanda:

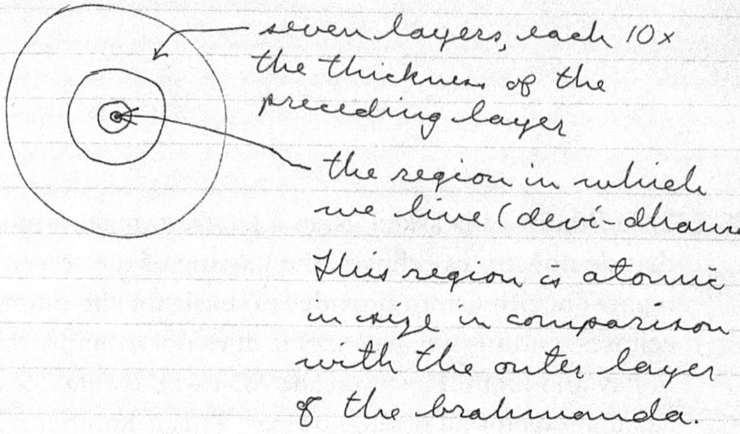

(c) devi-dhama:

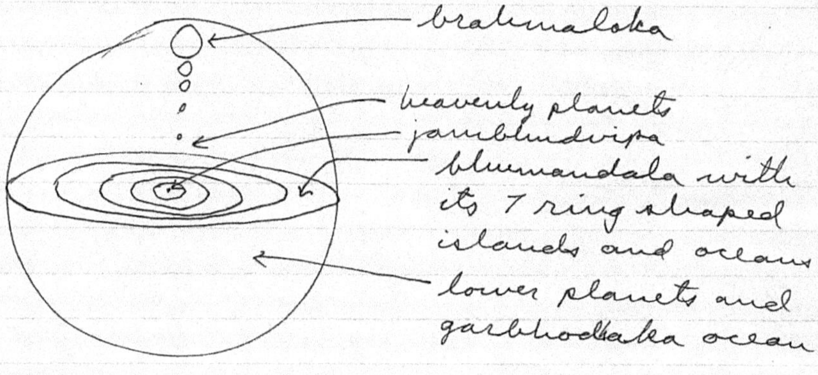

(d) the planetary system, consisting of sun, moon, mercury, venus, mars, jupiter and saturn, plus the stars: (not to scale)

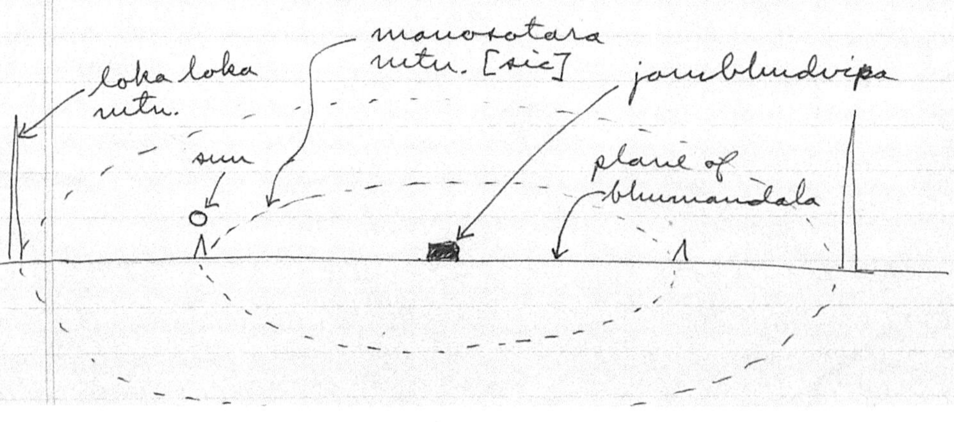

(b) one universal globe or *brahmāṇḍa*:

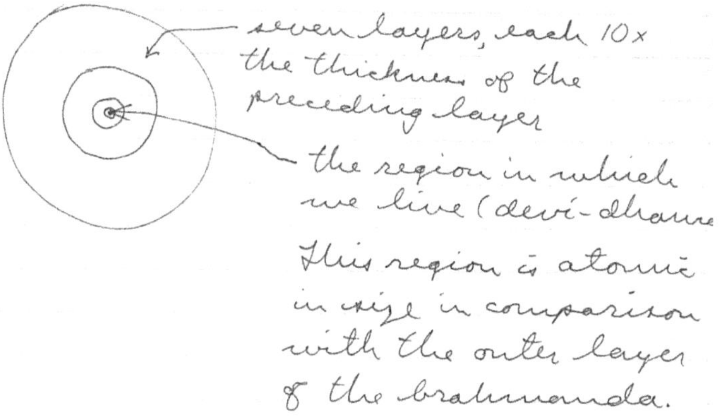

(c) Devī-dhāma:

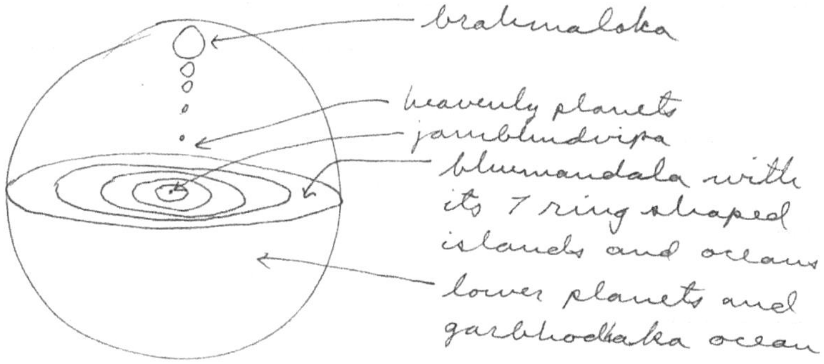

(d) the planetary system, consisting of the sun, the moon, Mercury, Venus, Mars, Jupiter and Saturn, plus the stars: (not to scale)

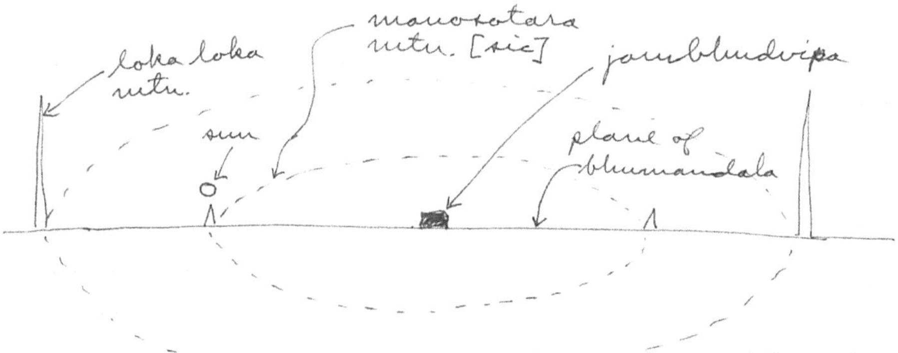

(6)

(e) Part (d), as seen from the side:

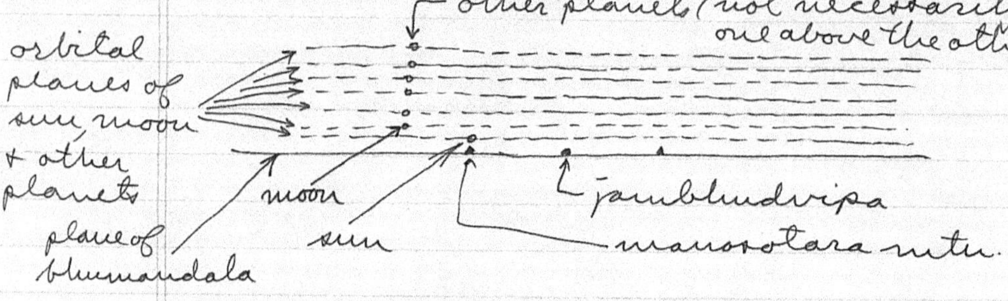

(f) A closer look at the sun and moon: Here are two possibilities.

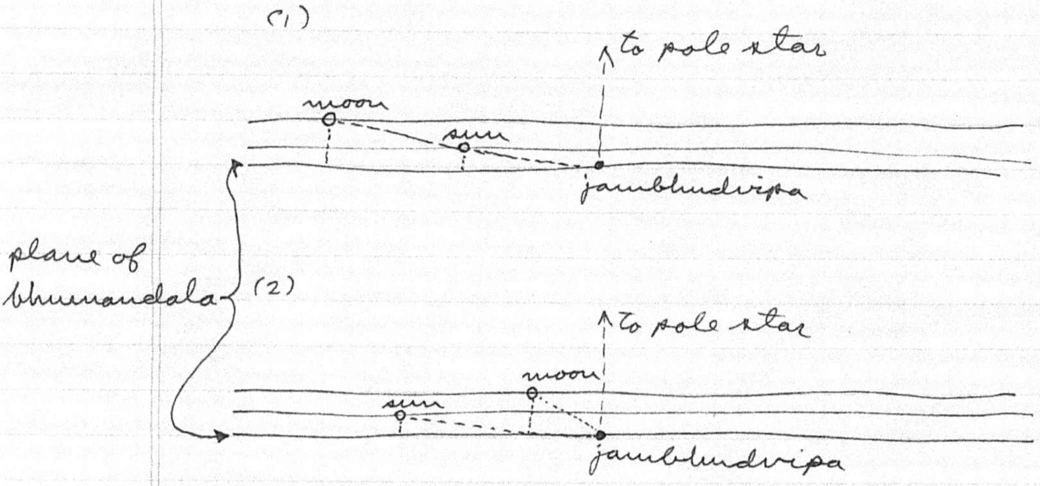

In both (1) and (2) we see that the moon is <u>higher</u> than the sun with respect to the plane of bhumandala. However, in (1) the moon is farther from jambhudvipa than the sun, whereas in (2) the moon is <u>closer</u> to jambhudvipa than the sun.

* Note, however, that in (2) the moon will appear to observers on jambhudvipa to be always closer (in angle) to the pole star than the sun. This is contrary to observation.

(e) Part (d), as seen from the side:

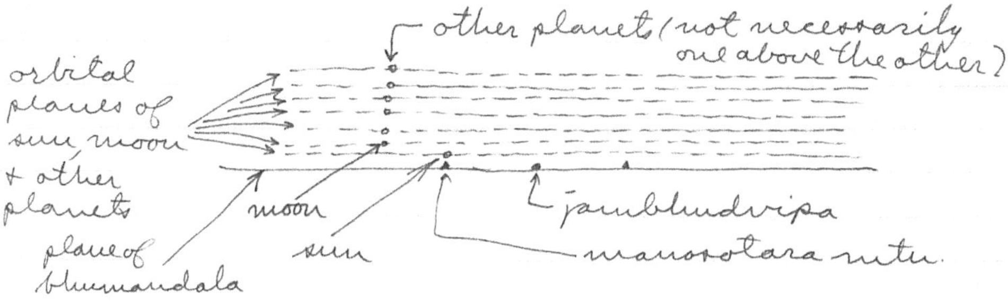

(f) A closer look at the sun and the moon – here are two possibilities:

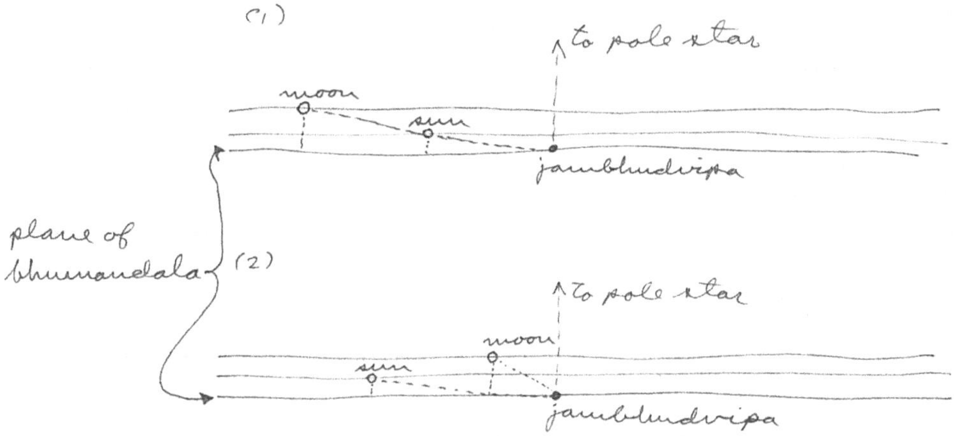

In both (1) and (2) we see that the moon is *higher* than the sun with respect to the plane of Bhū-maṇḍala. However, in (1) the moon is farther from Jambūdvīpa than the sun, whereas in (2) the moon is *closer* to Jambūdvīpa than the sun.

★ Note, however, that in (2) the moon will appear to observers on Jambūdvīpa to be always closer (in angle) to the pole star than the sun. This is contrary to observation.

(g) Jambhudvipa— the place in the universe where we live:

(1) map (looking from above):

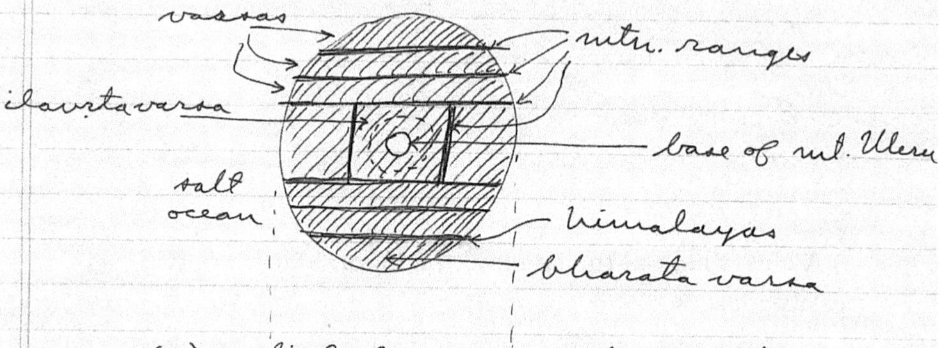

(2) relief from side (actually, jambhudvipa has been rotated by 90° for this view)

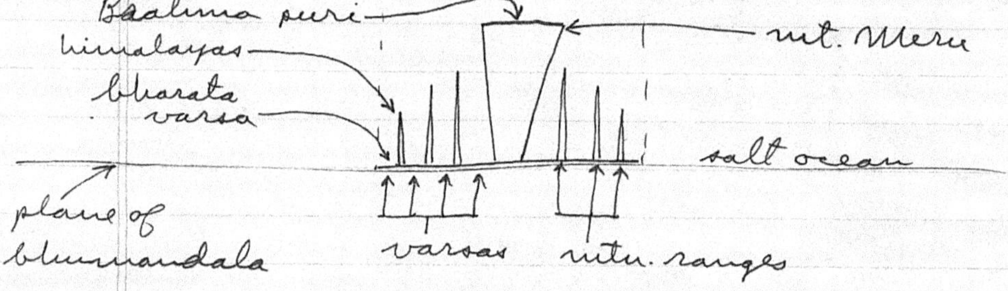

(3) another possibility, which takes into account the fact that the Ganges flows from the top of Mt. Meru to the salt ocean:

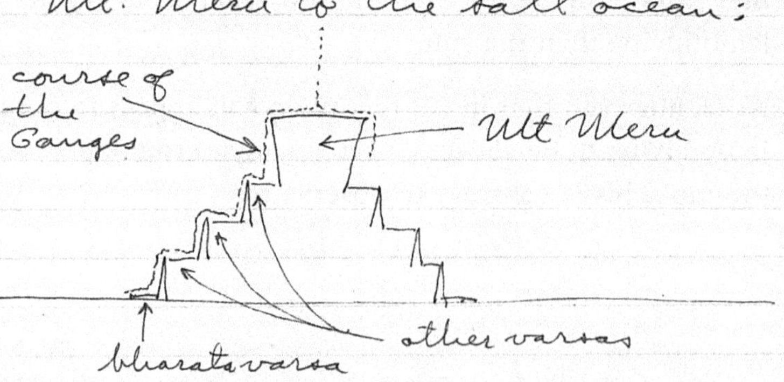

(g) Jambūdvīpa – the place in the universe where we live:

(1) map (looking from above):

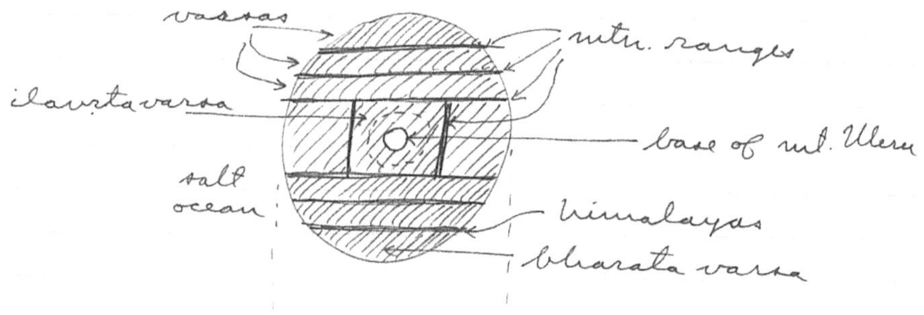

(2) relief from side (actually, Jambūdvīpa has been rotated by 90° for this view).

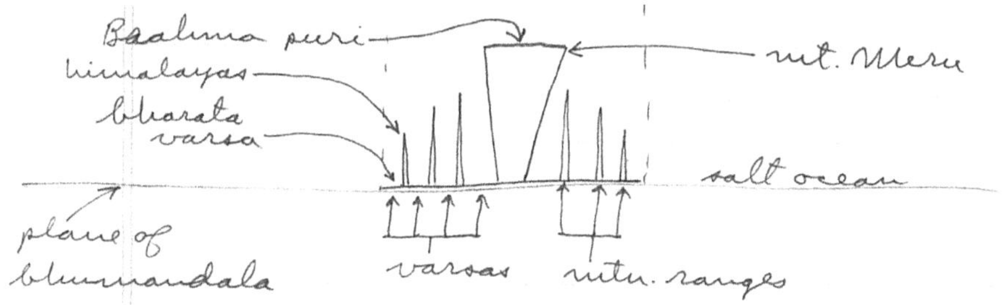

(3) another possibility, which takes into account the fact that the Ganges flows from the top of Mt. Meru to the salt ocean:

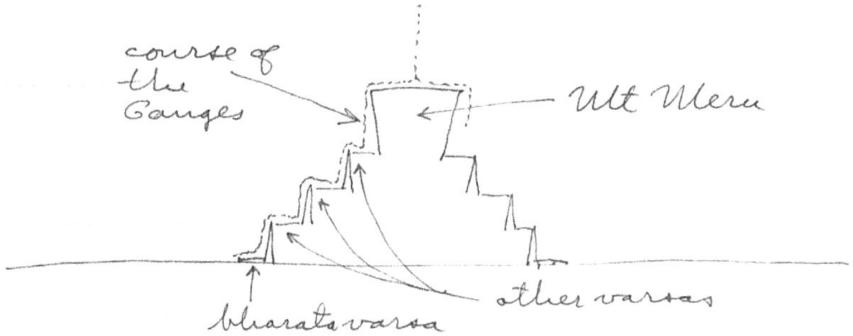

(4) poetic description of jambhudvipa in S.B. as a lotus flower:

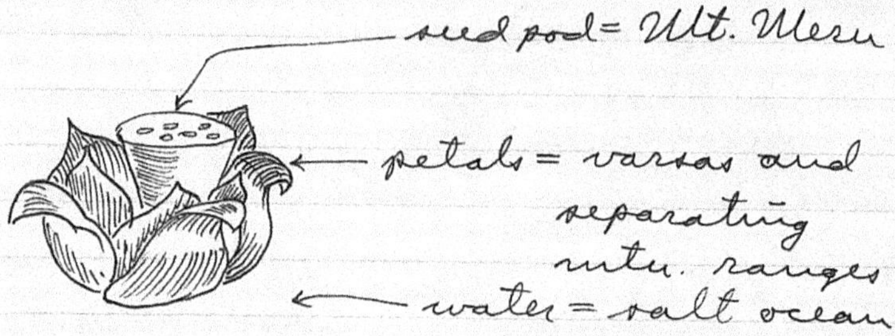

10. Question: Where is the earth planet, considered as a globe?
Answer: Here is a comparison, drawn roughly to scale, of the 7000 mile diameter earth globe (as we conceive of it) and jambhudvipa:

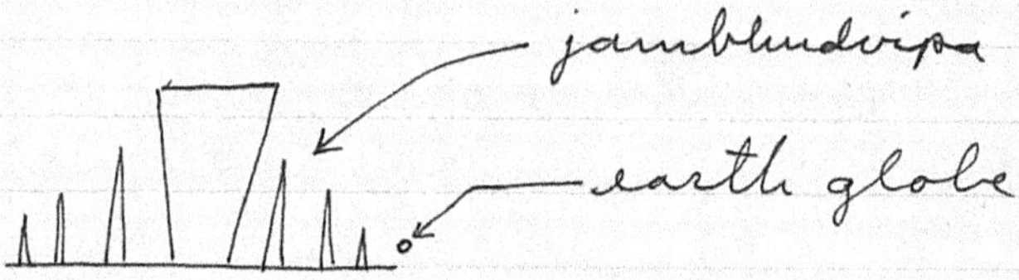

Comparison of earth globe and bharata varsa:

11. Can the earth globe be part of jambhudvipa?
(a) one possibility (the pie a la mode model:

(4) poetic description of Jambūdvīpa in *S.B.* as a lotus flower:

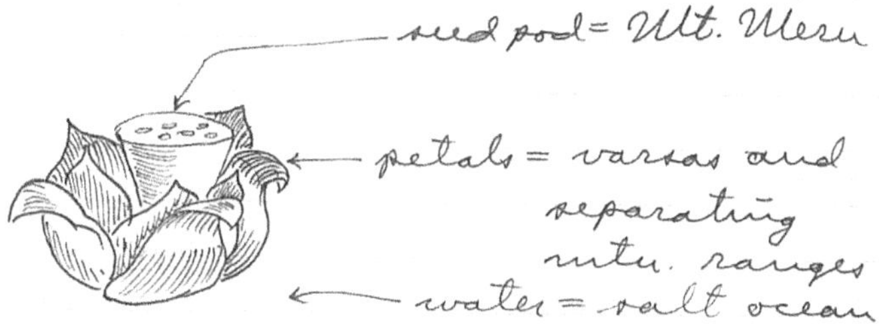

10. Question: Where is the earth planet, considered as a globe?

Answer: Here is a comparison, drawn roughly to scale, of the 8,000-mile-diameter earth globe (as we conceive of it) and Jambūdvīpa:

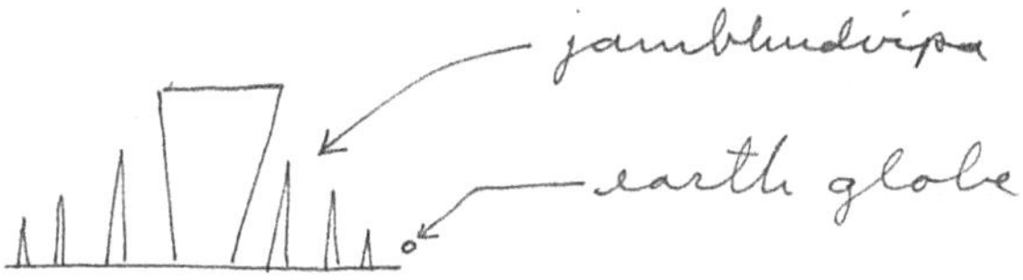

Comparison of earth globe and Bhārata-varṣa:

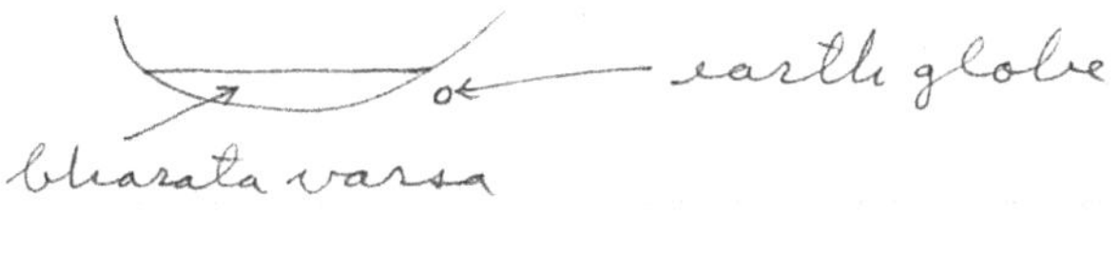

11. Can the earth globe be part of Jambūdvīpa?

(a) one possibility (the pie a la mode model):

(9)

N.B. The Chinese pie a la mode model.

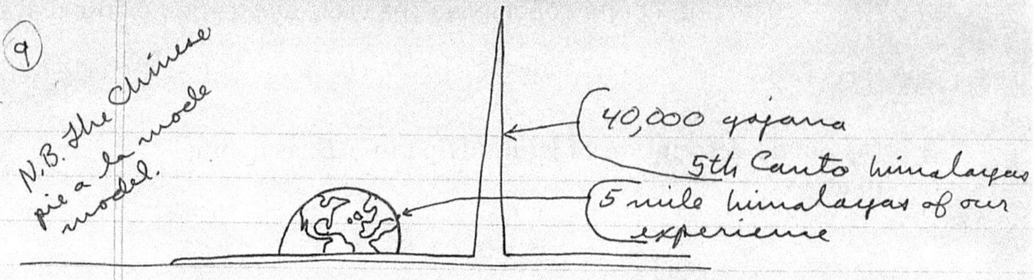

Problems: what happens at equator?
what became of the southern hemisphere? etc.

(b) another possibility (the golf tee model):

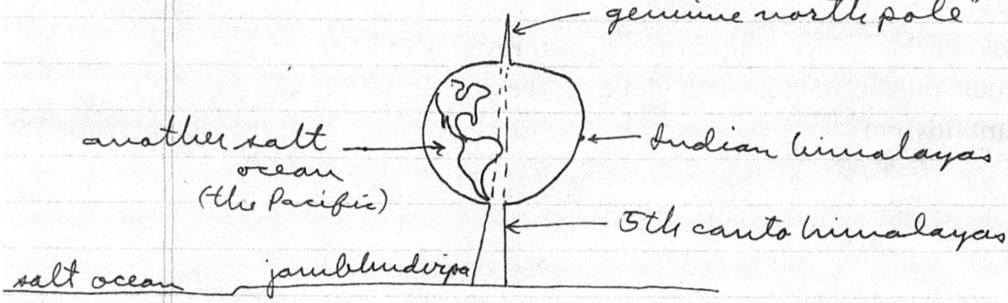

Problems: plenty.

(c) Yet another

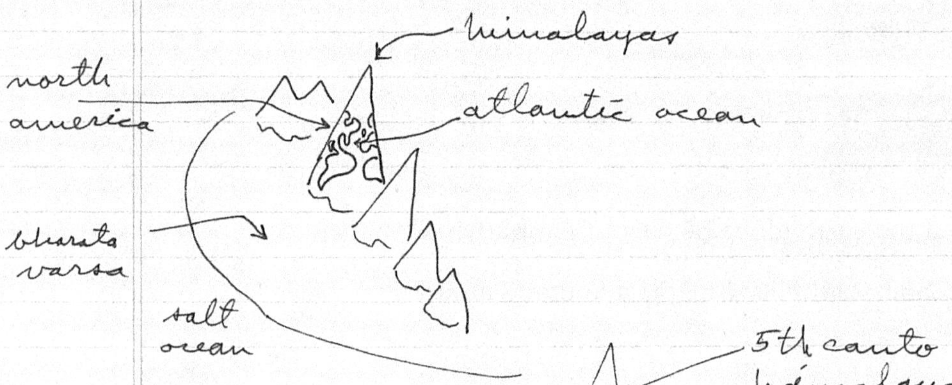

(d) and another

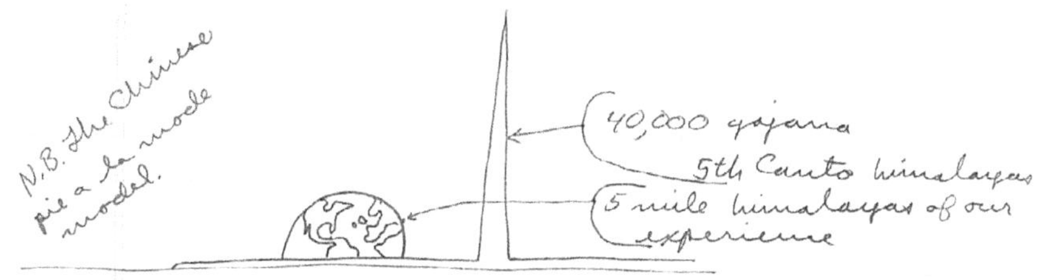

Problems: What happens at the equator? What became of the Southern Hemisphere? etc.

(b) another possibility (the golf tee model):

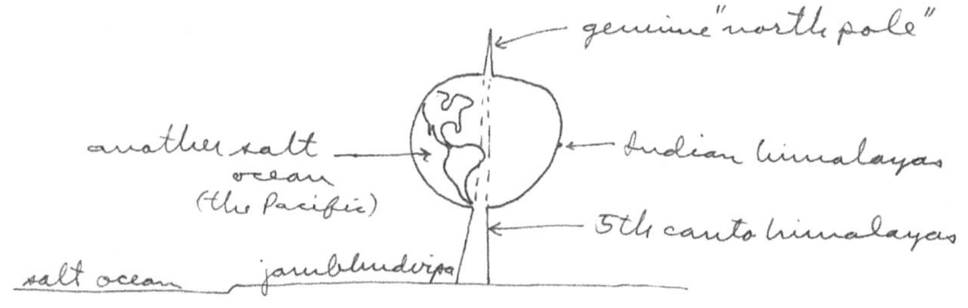

Problems: plenty.

(c) yet another

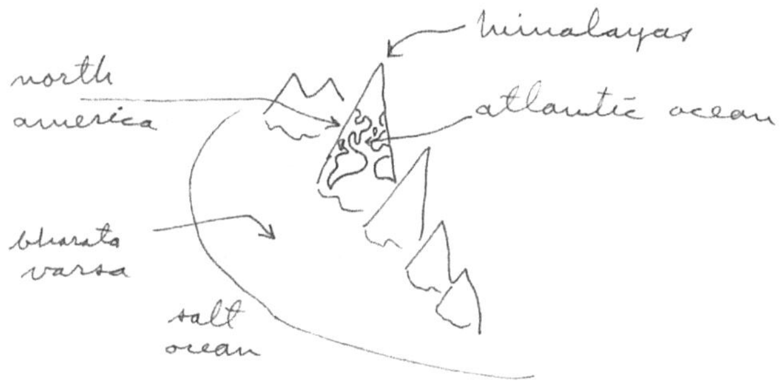

(d) and another

(e) There are many other models of this kind.

12. None of these models are at all plausible or defensible. They extensively contradict our ordinary experience; they are not supported by the *Bhagavatam* itself; and they give rise to more problems than they solve. (*Note*: Are any of these models supported in other Puranas, etc.?)

13. The idea of the earth as a (relatively) small globe is there in the *Surya-siddhanta*; this idea is not a creation of modern science, nor did it originate in Europe or ancient Greece. In fact, this idea was well known in ancient India.

* The conflict between the 5th Canto and our present idea of the earth globe is therefore not a conflict between modern science and Vedic knowledge. → It is an apparent conflict between two bodies of Vedic literature — the *Srimad Bhagavatam* (and other Puranas) and the *Surya siddhanta* (and other similar works).

14. The conflict between 5th Canto geography and the idea of the earth globe is more important than the contradiction involving the distances to the sun and moon. Unless we can resolve the former, there is no need to worry about the latter.

(e) There are many other models of this kind.

12. None of these models are at all plausible or defensible. They extensively contradict our ordinary experience; they are not supported by the *Bhāgavatam* itself; and they give rise to more problems than they solve. (NOTE: Are any of these models supported in other *Purāṇas*, etc.?)

13. The idea of the earth as a (relatively) small globe is there in the *Sūrya-siddhānta*; this idea is not a creation of modern science, nor did it originate in Europe or ancient Greece. In fact, this idea was well known in ancient India.

- ★ The conflict between the 5th Canto and our present idea of the earth globe is therefore not a conflict between modern science and Vedic knowledge. It is an apparent conflict between two bodies of Vedic literature – the *Śrīmad-Bhāgavatam* (and other *Purāṇas*) and the *Sūrya-siddhānta* (and other similar works).

14. The conflict between 5th Canto geography and the idea of the earth globe is more important than the contradiction involving the distances to the sun and moon. Unless we can resolve the former, there is no need to worry about the latter.

15. Here are various approaches to resolving this conflict:

(a) Fit the earth globe and jambhudvipa together geometrically in the same 3-dimensional space.

- This won't work, as we have seen.

(b) Deny that the earth is a globe. If we do this we must also reject the ~~5th Canto~~ *Surya-siddhanta*, which is said to be authentic by Srila Prabhupada. Also, we will have trouble explaining how it is that the devotees can travel all around the world in jet planes, following great circle routes.

(c) Reject the 5th Canto, claiming that it is the work of demons, or that it is allegorical. Here we face the problem that Srila Prabhupada translated and endorsed the 5th Canto, and never said that it was allegorical (or that it was the work of demons.)

16. The approaches in (15) are not acceptable. Let us also consider the approach of the scholars. They say that the 5th Canto records some ancient mythology that was conjured up by ignorant but poetic

15. Here are various approaches to resolving this conflict:

(a) Fit the earth globe and Jambūdvīpa together geometrically in the same 3-dimensional space – This won't work, as we have seen.

(b) Deny that the earth is a globe. If we do this, we must also reject the *Sūrya-siddhānta*, which is said to be authentic by Śrīla Prabhupāda. Also, we will have trouble explaining how it is that the devotees can travel all around the world in jet planes, following great circle routes.

(c) Reject the 5th Canto, claiming that it is the work of demons, or that it is allegorical. Here we face the problem that Śrīla Prabhupāda translated and endorsed the 5th Canto, and never said that it was allegorical (or that it was the work of demons.)

16. The approaches in (15) are not acceptable. Let us also consider the approach of the scholars. They say that the 5th Canto records some ancient mythology that was conjured up by ignorant but poetic sages who lived in primitive Aryan tribes. They say that the *Surya-siddhanta* represents a later, more sophisticated understanding of astronomy that was imported into India from Greece and then assimilated into "Hindu" culture. (Note: *S.S.* places Mt. Meru at the North pole. The scholars say that this is an example of compromise between the Greek and Puranic cosmology.)

Here is one argument in opposition to this interpretation: "Primitive" people are generally more sophisticated in their understanding of astronomical phenomena than the typical modern Western city dweller. For example, the wise men of the Maoris of N.Z. had names for many stars, and they knew when they would first rise and set during the year. Presumably the ancient Vedic sages, however "primitive" they may have been, would have had comparable astronomical sophistication.

Now, it so happens that the most direct interpretation (in terms of our customary framework of thought) of the 5th Canto does not enable us to understand how the sun rises in the East, proceeds to a high point in the sky, and then sets in the West. The following diagram shows why this is so:

sages who lived in primitive Aryan tribes. They say that the *Sūrya-siddhānta* represents a later, more sophisticated understanding of astronomy that was imported into India from Greece and then assimilated into "Hindu" culture. (Note: *S.S.* places Mt. Meru at the north pole. The scholars say that this is an example of compromise between the Greek and Puranic cosmology.)

Here is one argument in opposition to this interpretation: "Primitive" people are generally more sophisticated in their understanding of astronomical phenomena than the typical modern Western city dweller. For example, the wise men of the Māoris of New Zealand had names for many stars, and they knew when they would first rise and set during the year. Presumably the ancient Vedic sages, however "primitive" they may have been, would have had comparable astronomical sophistication.

Now, it so happens that the most direct interpretation (in terms of our customary framework of thought) of the 5th Canto does not enable us to understand how the sun rises in the East, proceeds to a high point in the sky, and then sets in the West. The following diagram shows why this is so:

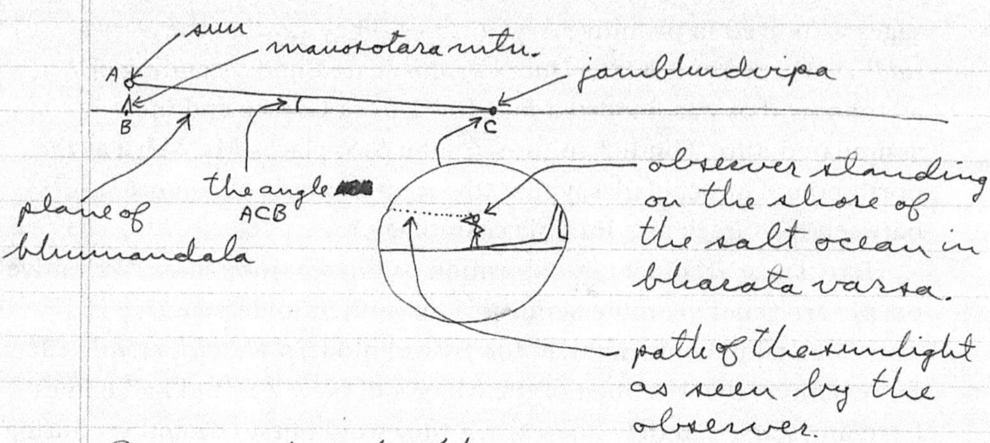

According to the 5th Canto, the distance AB from the sun to the plane of bhumandala (at the base of manosotara mtn.) is much smaller than the distance BC from manosotara mtn. to jambhudvipa. Therefore the angle ACB between the sun and the plane of bhumandala must always be very small. Now, the salt ocean lies in the plane of bhumandala. Hence an observer standing by the shore of the salt ocean will never see the sun rise very high above the horizon. Indeed, the sun will skim the horizon in the same way that it does in the "land of the midnight sun" in the arctic.

One might say, "If the land on which we stand is tilted with respect to bhumandala, as in 11(a), (b), or (c), then it may be possible to see an ordinary sunrise and sunset. However, the Bhagavatam gives no direct indication that 11(a), (b), or (c) are true, and there is also another

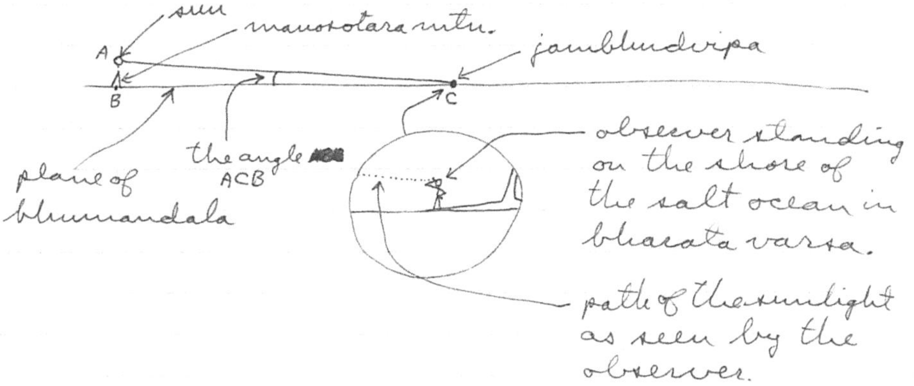

According to the 5th Canto, the distance AB from the sun to the plane of Bhū-maṇḍala (at the base of Mānasottara Mountain) is much smaller than the distance BC from Mānasottara Mountain to Jambūdvīpa. Therefore, the angle ACB between the sun and the plane of Bhū-maṇḍala must always be very small. Now, the salt ocean lies in the plane of Bhū-maṇḍala. Hence an observer standing by the shore of the salt ocean will never see the sun rise very high above the horizon. Indeed, the sun will skim the horizon in the same way that it does in the "land of the midnight sun" in the arctic.

One might say, "If the land on which we stand is tilted with respect to Bhū-maṇḍala, as in 11(a), (b), or (c), then it may be possible to see an ordinary sunrise and sunset. However, the *Bhāgavatam* gives no direct indication that 11(a), (b), or (c) are true, and there is also another problem: (which we illustrate using 11(b)).

problem: (which we illustrate using 11(b).)

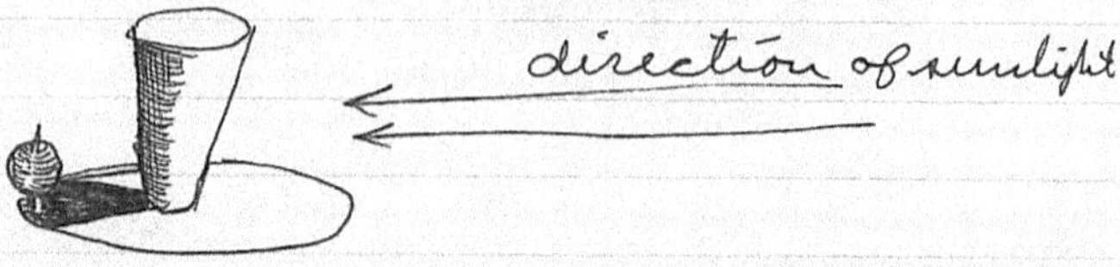

This is, that the side of the earth facing Mt. Meru will experience darkness at noon every day.
Of course, one could try

or

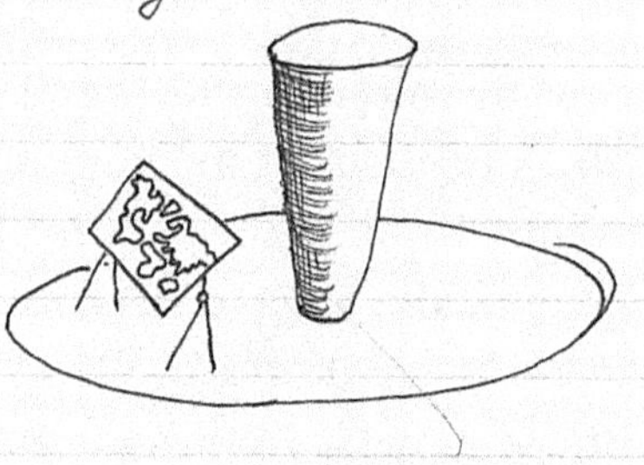

but there are problems with these models also.

17. The point of (16) is that the 5th Canto, interpreted literally as a 3-d. description of the universe, could not have made sense even to the most naive of "primitive Vedic sages."
Therefore, Vyasadeva must have had something else in mind.

18. We suggest that Vyasadeva must have been conceiving of the universe in terms of ideas of space (and time) that are quite different from our own.

19. First of all note that even ideas such as 11 (a), (b) + (c) + (d) require some sort of distortion of space

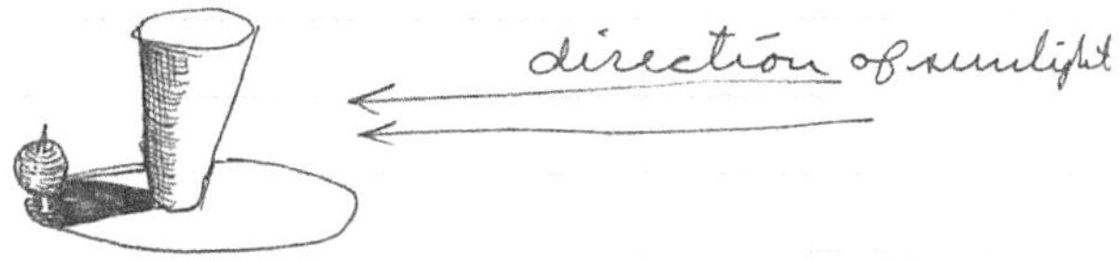

This is, that the side of the earth facing Mt. Meru will experience darkness at noon every day. Of course, one could try:

but there are problems with these models also.

17. The point of (16) is that the 5th Canto, interpreted literally as a 3-D description of the universe, could not have made sense even to the most naive of "primitive Vedic sages."

Therefore, Vyāsadeva must have had something else in mind.

18. We suggest that Vyāsadeva must have been conceiving of the universe in terms of ideas of space (and time) that are quite different from our own.

19. First of all note that even ideas such as 11(a), (b), (c), and (d) require some sort of distortion of space.

(15)

If they are to work: 11(a)+(b)+(c) require some spatial "magic" at the join between the earth globe and jambhudvipa. 11(d) also requires this, and it also requires some geometrical "force" or "warp" that makes great circle routes shortest on a flat surface.

20. If we try to change the size of the yojana and make jambhudvipa comparable in size with the earth, then some sophisticated spatial transformations are still needed to make it comparable in shape.
(a) Note the earth model given in the Siddhanta-siromani (which Srila Prabhupada cites for its reference to the seven dvipas):

Could this be realistic, or is it a wild speculation by the author of Siddhanta-siromani? If it is realistic, we have to map the southern hemisphere onto the plane of bhumandala by some kind of sophisticated transformation (and what does this transformation do to

We also have to relate the 7 dvipas of Siddhanta siromani with Africa, South America, Australia, and Antarctica.

If they are to work: 11(a) and (b) and (c) require some spatial "magic" at the join between the earth globe and Jambūdvīpa. 11(d) also requires this, and it also requires some geometrical "force" or "warp" that makes great circle routes shortest on a flat surface.

20. If we try to change the size of the *yojana* and make Jambūdvīpa comparable in size with the earth, then some sophisticated spatial transformations are still needed to make it comparable in *shape*.

(a) Note the earth model given in the *Siddhānta-śiromaṇi* (which Śrīla Prabhupāda cites for its reference to the seven *dvīpas*):

Could this be realistic, or is it a wild speculation by the author of *Siddhānta-śiromaṇi*? If it is realistic, we have to map the Southern Hemisphere onto the plane of Bhū-maṇḍala by some kind of sophisticated transformation (and what does this transformation do to the shell of the universe, which in the 5th Canto meets the outer edge of the Bhū-maṇḍala plane?)

[Side note] We also have to relate the 7 *dvīpas* of *Siddhānta-śiromaṇi* with Africa, South America, Australia, Antarctica [etc.].

the shell of the universe, which in the 5th Canto meets the outer edge of the bhumandala plane?)

(b) Note that in some purports of the S.B., Srila Prabhupada has identified parts of the earth as we know it with certain varsas in Jambhudvipa. There is also a purport identifying the dvipas with the continents of the earth as we know it. These identifications entail some kind of sophisticated spatial transformation between the 5th Canto universe, directly interpreted, and the earth of our experience.

21. Just for the fun of it, here is a transformation mapping a plane to the surface of a sphere:

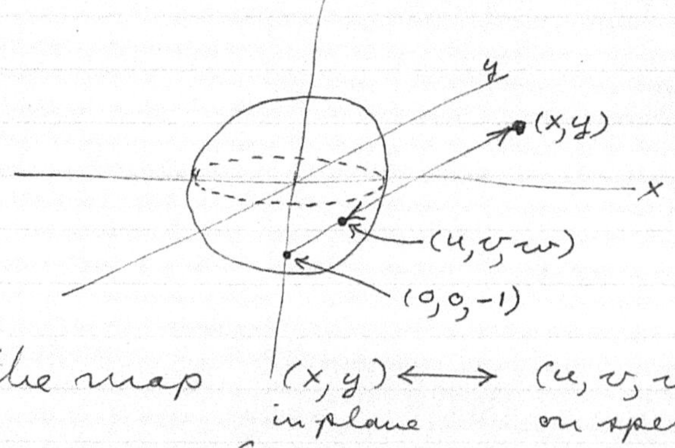

Here the map $(x,y) \longleftrightarrow (u,v,w)$
in plane — on sphere

is given by:

$$\begin{cases} u = \dfrac{2x}{x^2+y^2+1} \qquad v = \dfrac{2y}{x^2+y^2+1} \\ w = \dfrac{2}{x^2+y^2+1} - 1 \end{cases}$$

(b) Note that in some purports of the *S.B.*, Śrīla Prabhupāda has identified parts of the earth as we know it with certain *varṣas* in Jambūdvīpa. There is also a purport identifying the *dvīpas* with the continents of the earth as we know it. These identifications entail some kind of sophisticated spatial transformation between the 5th Canto universe, directly interpreted, and the earth of our experience.

21. Just for the fun of it, here is a transformation mapping a plane to the surface of a sphere:

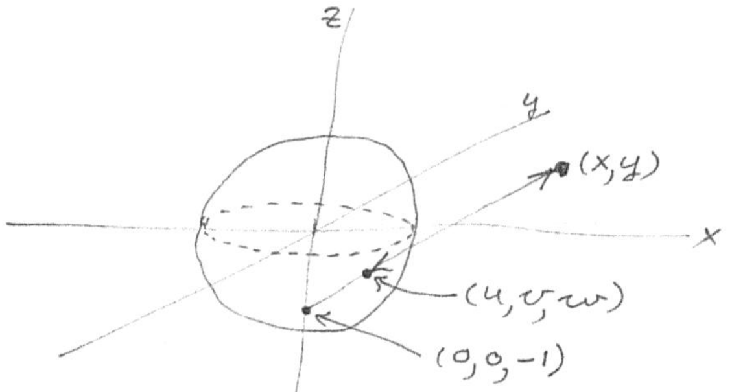

Here the map $(x, y) \longleftrightarrow (u, v, w)$
in plane on sphere

is given by:

$$\begin{cases} u = \dfrac{2x}{x^2+y^2+1} \qquad v = \dfrac{2y}{x^2+y^2+1} \\ w = \dfrac{2}{x^2+y^2+1} - 1 \end{cases}$$

(Note: This maps far regions of the plane to points near the south pole of the sphere, as required for the *Siddhānta-śiromaṇi* model.)

(11)

(Note: This maps far regions of the plane to points near the south pole of the sphere, as required for the Siddhānta-śiromaṇi model.) It is not at all clear what it would mean to say that bhūmaṇḍala and the earth globe are related by such a transformation, and we think that it is fruitless to pursue the study of such transformations in our effort to understand the 5th Canto.

22. A more sophisticated idea is necessary. We suggest that such an idea can be obtained by directly examining what the S.B. has to say about the nature of space, and then explicating these ideas (to the extent that this is possible) within the framework of the mathematical field known as topology.

23. We will derive the following basic principles: space is determined by modes of travel. The S.B. refers to modes of travel that are quite different from the modes of travel familiar to us (but which include them.) Thus the S.B. refers to a idea of space which goes beyond our customary idea of 3-dimensional space (but which includes it.)

We propose that reality as we experience is a subset of the total reality being described in the 5th Canto description of the universe, and that this subset is determined by the limited modes of travel and perception available to us. We propose

It is not at all clear what it would mean to say that Bhū-maṇḍala and the earth globe are related by such a transformation, and we think that it is fruitless to pursue the study of such transformations in our effort to understand the 5th Canto.

22. A more sophisticated idea is necessary. We suggest that such an idea can be obtained by directly examining what the *S.B.* has to say about the nature of space, and then explicating these ideas (to the extent that this is possible) within the framework of the mathematical field known as topology.

23. We will derive the following basic principles: Space is determined by modes of travel. The *S.B.* refers to modes of travel that are quite different from the modes of travel familiar to us (but which include them). Thus the *S.B.* refers to an idea of space which goes beyond our customary idea of 3-dimensional space (but which includes it).

We propose that reality as we experience it is a subset of the total reality being described in the 5th Canto description of the universe, and that this subset is determined by the limited modes of travel and perception available to us. We propose that the 5th Canto is presenting a map of the universe as it is seen by yogis who can perceive and travel using yogic siddhis.

S.S. + eclipses — planetarium either no spatial thinking — or a more soph. kind.

24. We begin with some ideas about space.

(a) Historical. In various cultures people have had various conceptions of space, many of which are quite different from our own.

* For many centuries the idea of Euclidean geometry dominated European thinking. 3-D Euclidean space was used by Newton as the basis for his physics. He thought of this as "absolute space", and later on, Kant tried to prove that 3-D Euclidean space is absolute.
* In the 19th(?) century Lobachevsky [sic], Bolyai, Gauss, and Reimann developed the idea of non-Euclidean geometry.

We can visualize various 2-D spaces embedded within 3-D Euclidean space:

(1) Euclidean 2-D plane:

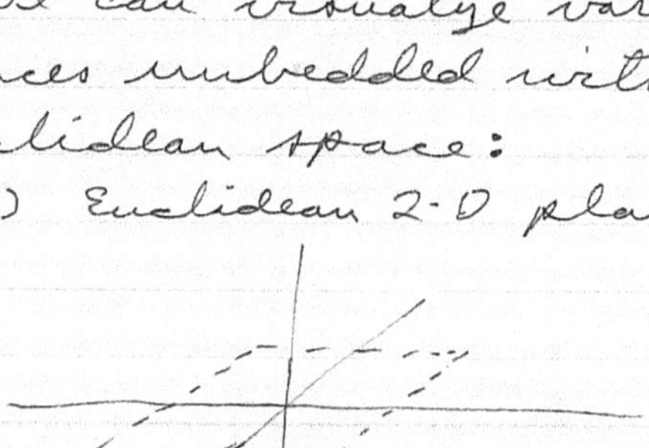

that the 5th Canto is presenting a map of the universe as it is seen by yogis who can perceive and travel using yogic *siddhis*.

24. We begin with some ideas about space.

(a) Historical. In various cultures people have had various conceptions of space, many of which are quite different from our own.

- ⋆ For many centuries, the ideas of Euclidean geometry dominated European thinking. 3-D Euclidean space was used by Newton as the basis for his physics. He thought of this as "absolute space," and later on, Kant tried to prove that 3-D Euclidean space is absolute.
- ⋆ In the 19th century Lobachevsky, Bolyai, Gauss, and Riemann developed the idea of non-Euclidean geometry.

We can visualize various 2-D spaces embedded within 3-D Euclidean space:

(1) Euclidean 2-D plane:

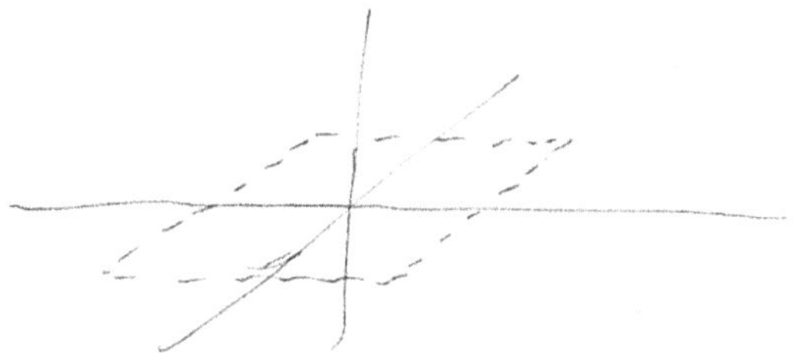

[Side note] S.S. and eclipses – planetarium: either no spatial thinking, or a more sophisticated kind.

(2) non-Euclidean 2-D spherical surface:

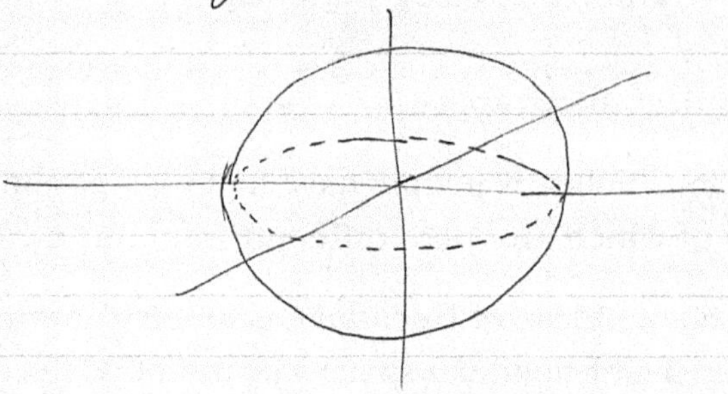

(3) non-Euclidean 2-D hyperbolic surface:

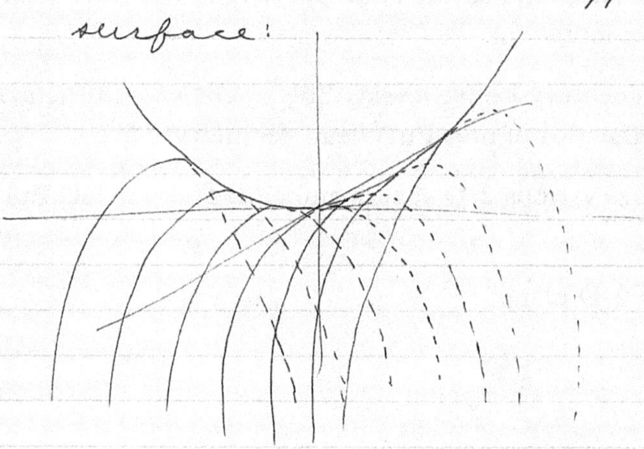

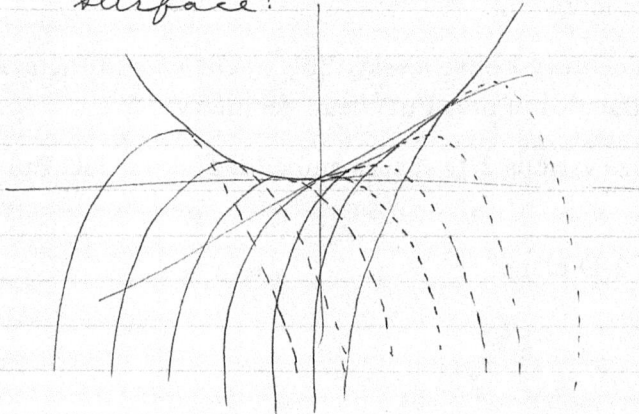

In (1) a geodesic, or straightest line, is an ordinary straight line; in (2) it is a great circle; and in (3) it is another kind of curve. The points and geodesics in (1), (2) & (3) satisfy all of the Euclidean axioms except for the parallel postulate.

Given a line and a pt. not on the line: In (1) there is one line through the pt. that misses the line.

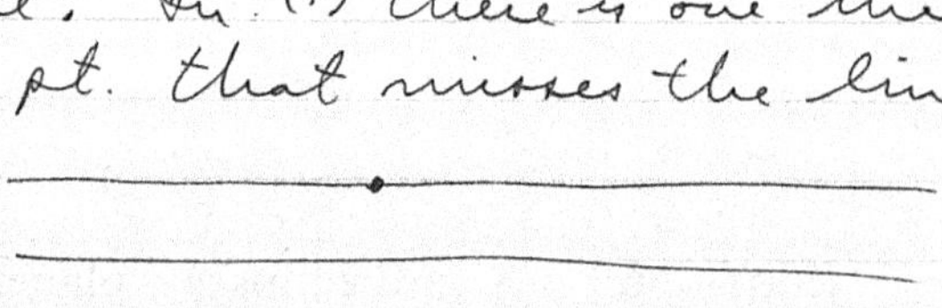

Thus in (1) the parallel postulate holds.

(2) non-Euclidean 2-D spherical surface:

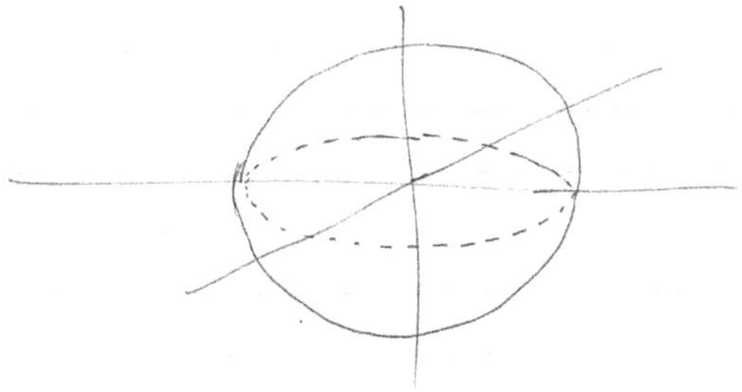

(3) non-Euclidean 2-D hyperbolic surface:

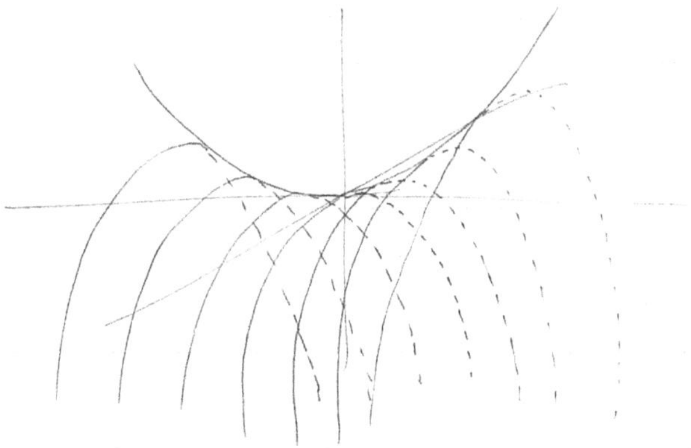

In (1) a geodesic, or straightest line, is an ordinary straight line; in (2) it is a great circle; and in (3) it is another kind of curve. The points and geodesics in (1), (2), and (3) satisfy all of the Euclidean axioms except for the parallel postulate.

Given a line and a point not on the line: In (1) there is one line through the point that misses the line.

Thus in (1) the parallel postulate holds.

In (2) every line (geodesic) through the pt. intersects the given line:

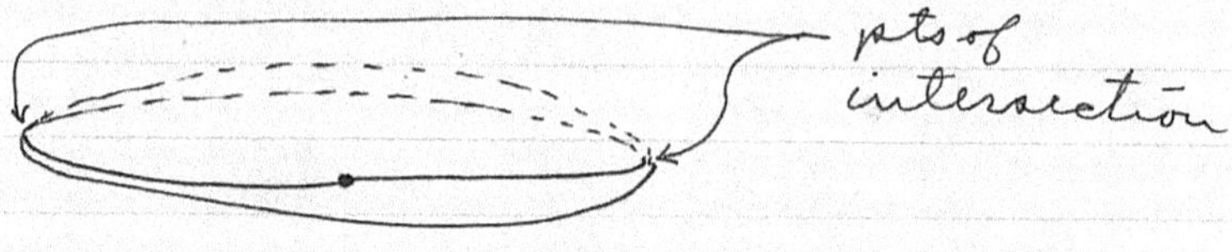

In (3) there are many lines (geodesics) through the pt. which miss the given line:

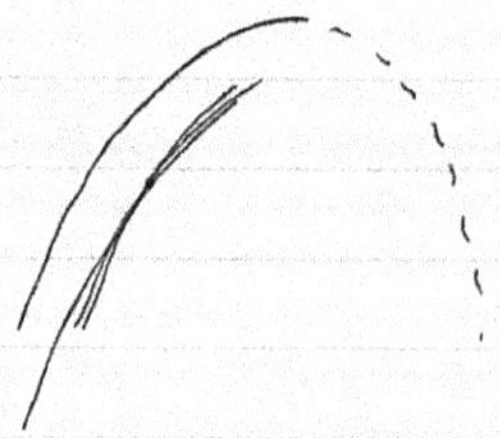

* Spaces (1), (2) + (3) are embedded in a higher dimensional Euclidean space, but they can be described without reference to such a higher space.

* The pythagorean theorem is violated in (2) and (3):

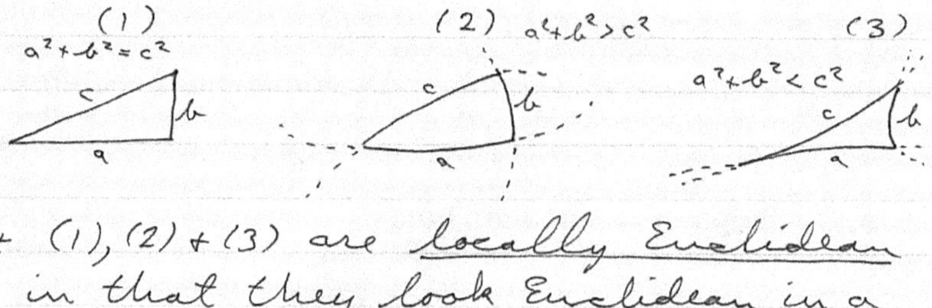

* (1), (2) + (3) are <u>locally Euclidean</u> in that they look Euclidean in a sufficiently small region around any point. (i.e. a small region on the surface of a sphere looks flat.)

In (2) every line (geodesic) through the point intersects the given line:

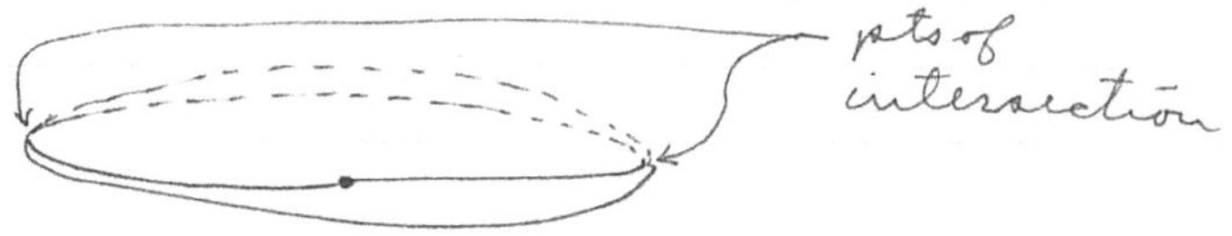

In (3) there are many lines (geodesics) through the point which miss the given line:

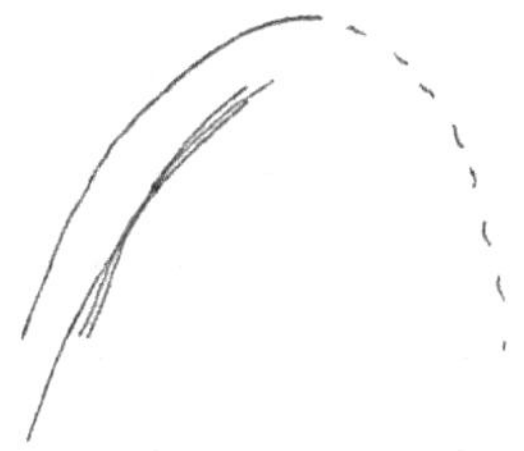

* Spaces (1), (2) and (3) are embedded in a higher-dimensional Euclidean space, but they can be described without reference to such a higher space.
* The Pythagorean theorem is violated in (2) and (3):

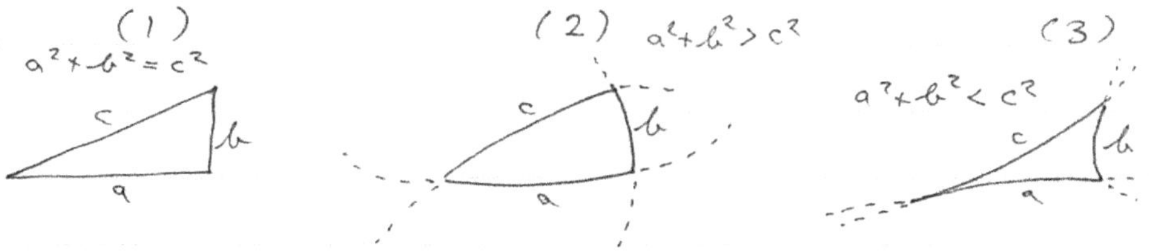

* (1), (2) and (3) are locally Euclidean in that they look Euclidean in a sufficiently small region around any point (i.e., a small region on the surface of a sphere looks flat).

* Reimann found a very general way of describing non-Euclidean spaces in terms of curvilinear coordinates.

* An example of 3-D non-Euclidean space: the "surface" of a 4-D sphere. This is hard to draw but easy to define in terms of coordinates: It is the set of (x, y, z, w) such that $x^2+y^2+z^2+w^2=r^2$ (r = radius of sphere.)

* Man looking at himself from behind in a small 3-D space of this kind:

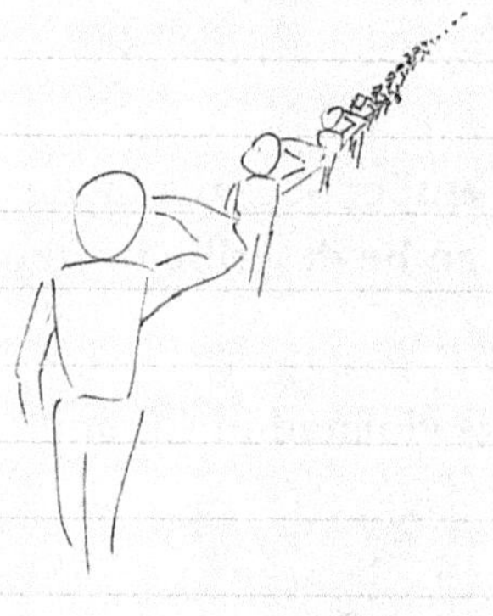

* The Moebius strip, in which an object, after traversing a closed loop, becomes reversed from right to left:

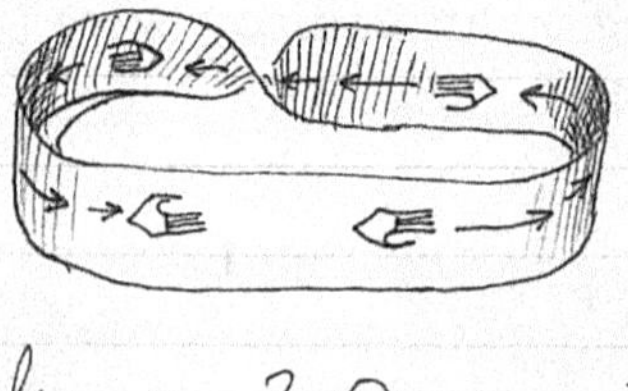

In a 3-D version of this a person following the closed loop would change from right handed to left handed.

★ Riemann found a very general way of describing non-Euclidean spaces in terms of curvilinear coordinates.

★ An example of 3-D non-Euclidean space: the "surface" of a 4-D sphere. This is hard to draw but easy to define in terms of coordinates: It is the set of (x, y, z, w) such that $x^2 + y^2 + z^2 + w^2 = r^2$ (r = radius of sphere).

★ Man looking at himself from behind in a small 3-D space of this kind:

★ The Möbius strip, in which an object, after traversing a closed loop, becomes reversed from right to left:

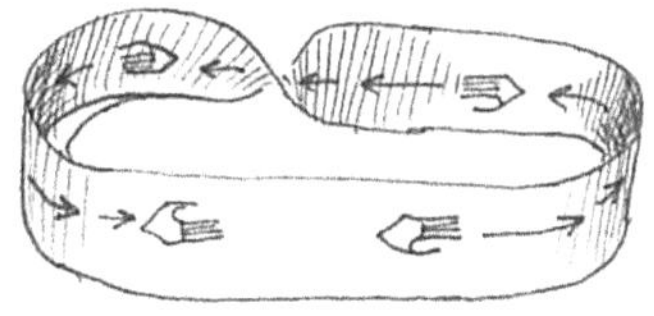

In a 3-D version of this, a person following the closed loop would change from right handed to left handed.

(22)

* A general theory of <u>topology</u> was devised for the purpose of systematically studying and classifying different kinds of space.

Basic principle:

Space = set of points (locations)
+
Neighborhood relations between points.

maps & U.S. subway maps

Illustration:

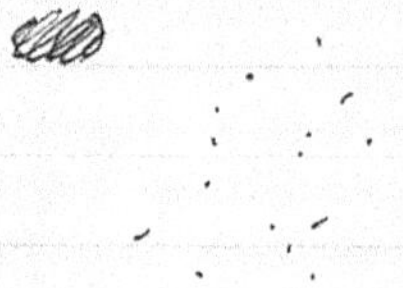

points with no nbd. relations are just isolated pts. (Do not visualize them as sitting within 3-D space, for space is what we are trying to define.)

Now let us add nbd. relations:

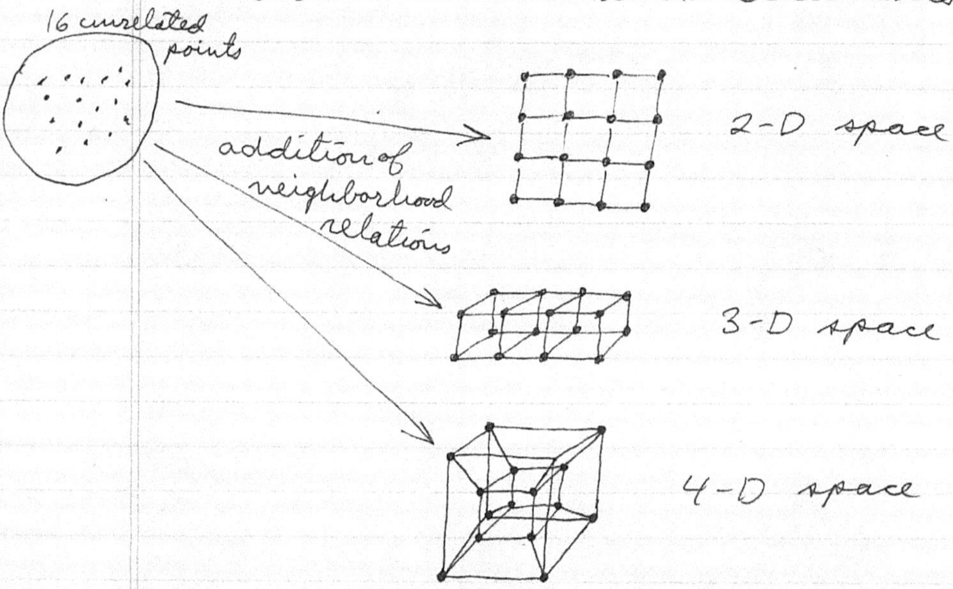

Thus, different spaces are defined by different rules governing which points

★ A general theory of topology was devised for the purpose of systematically studying and classifying different kinds of space.

Basic principle:

> Space = Set of points (locations) + Neighborhood relations between points.

Illustration:

Points with no neighborhood relations are just isolated points. (Do not visualize them as sitting within 3-D space, for space is what we are trying to define.)

Now let us add neighborhood relations:

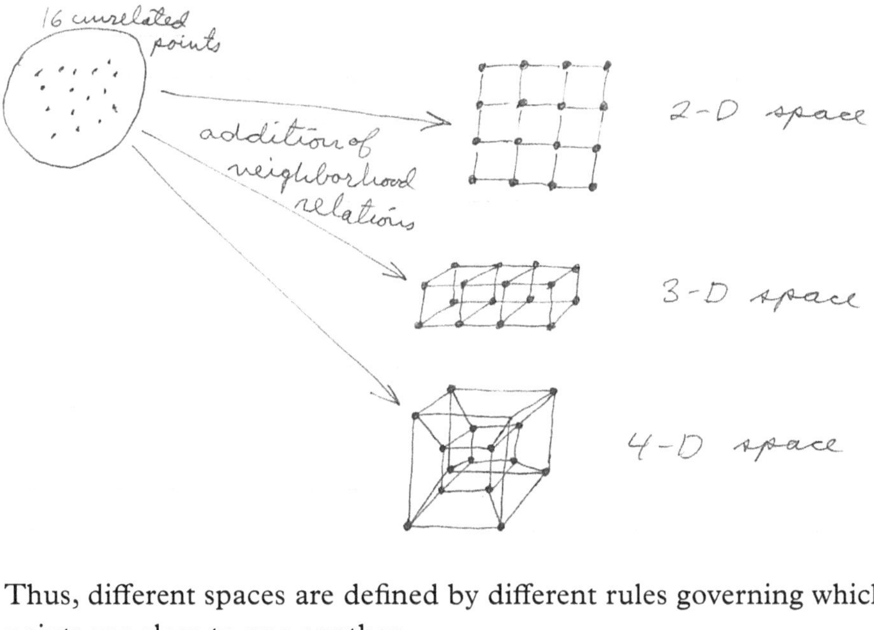

Thus, different spaces are defined by different rules governing which points are close to one another.

[Side note] maps of U.S. subway maps

are close to one another.

* In topology one can study:
  n-dimensional spaces (n = 1, 2, 3, ...)
  ∞-dimensional spaces
  spaces for which dimensionality has no meaning

* Now, what is the practical meaning of neighborhood relations? Well, two points are close to one another if travel from one to the other is easy, or if messages can be easily sent from one to the other. Otherwise they are effectively far apart.

(a) Forms of communication between points in ordinary experience:

- travel:

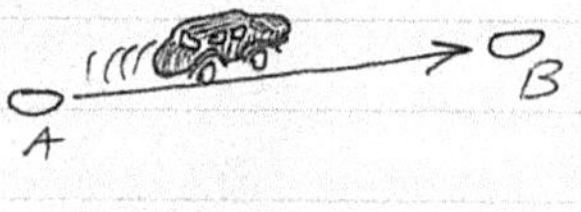

- wave propagation

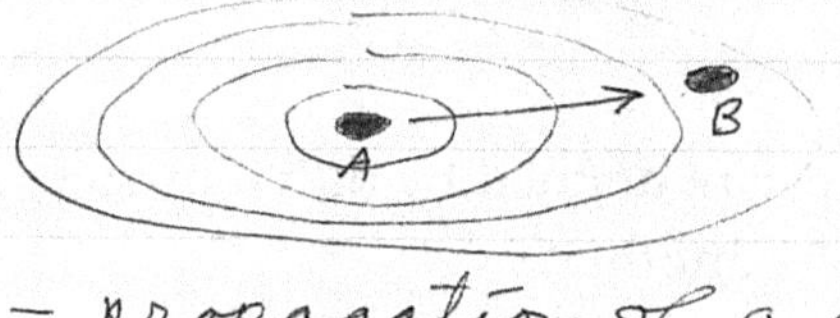

- propagation of a disturbance in other ways:

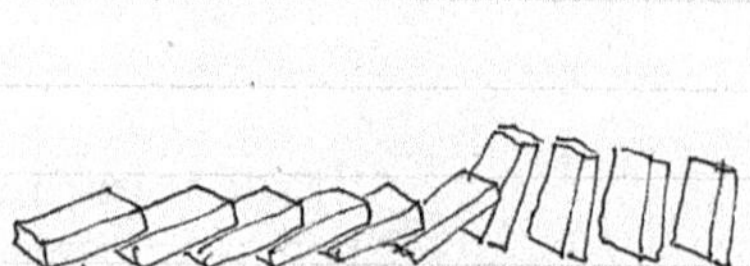

★ In topology one can study:

n-dimensional spaces (n = 1,2,3,...)

∞-dimensional spaces

Spaces for which dimensionality has no meaning

★ Now, what is the practical meaning of neighborhood relations? Well, two points are close to one another if travel from one to the other is easy, or if messages can be easily sent from one to the other. Otherwise they are effectively far apart.

(a) Forms of communication between points in ordinary experience:

– Travel:

– Wave propagation

– Propagation of a disturbance in other ways:

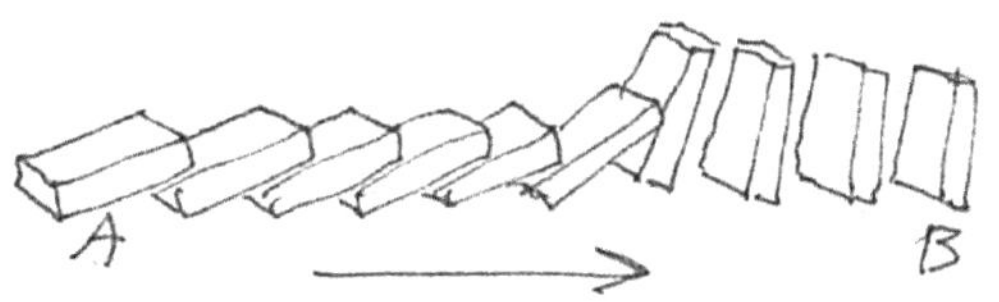

space created subway maps

We normally think of these modes of communication as occurring <u>in space</u>. Here we suggest that:

modes of communication ⇒ neighborhood relations ⇒ space

Thus modes of communication define space.

Examples:

(1) Suppose that a room has been divided by a perfectly impervious partition which will not allow any communication from one part of the room to the other. (Of course all partitions of our experience are not like this.) Then one part of the room has thereby become distant from the other part.

This kind of change in distance can be explained as follows in terms of nbd. relations:

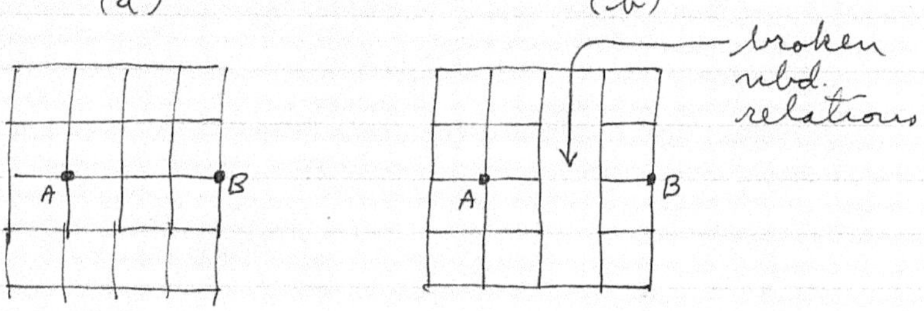

In (a) the distance from A to B is 3 links.
In (b) it is 7 links.

(2) Suppose that the map of country X looks like this:

We normally think of these modes of communication as occurring *in space*. Here we suggest that:

modes of communication → neighborhood relations → space

Thus modes of communication define space.

Examples:

(1) Suppose that a room has been divided by a perfectly impervious partition which will not allow any communication from one part of the room to the other. (Of course all partitions of our experience are not like this.) Then one part of the room has thereby become distant from the other part.

This kind of change in distance can be explained as follows in neighborhood relations:

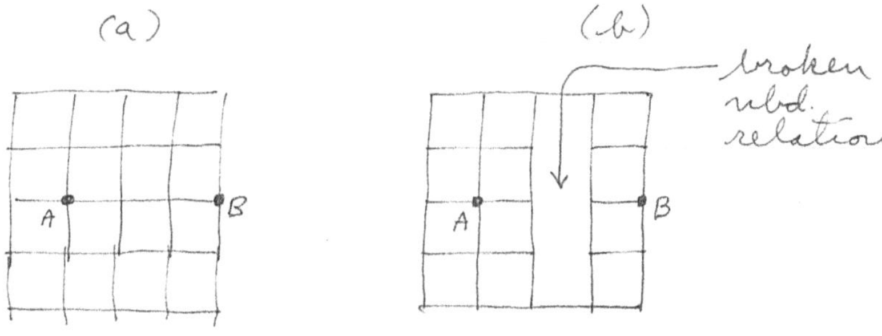

In (a) the distance from A to B is 3 links. In (b) it is 7 links.

(2) Suppose that the map of country X looks like this:

[Side note] space-created subway maps

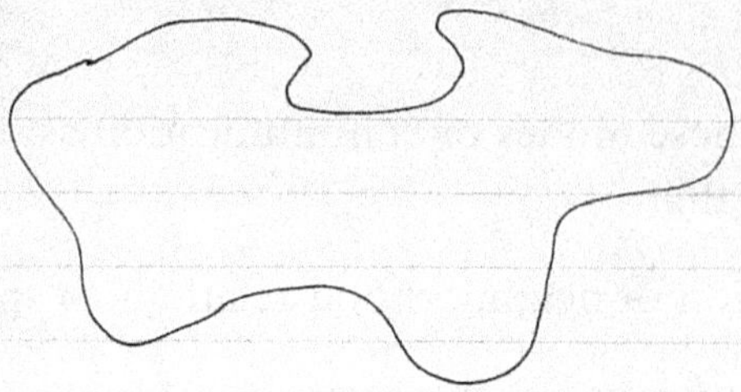

Now suppose that efficient air travel from north to south has become available, but that travel from east to west is only possible by land. Then, in effect, the map might be said to have shrunk in the vertical dimension:

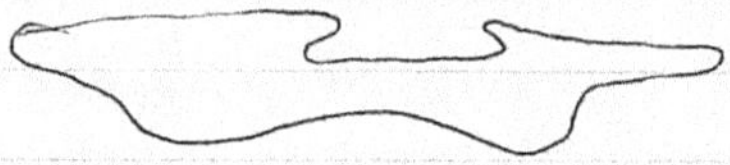

space diff for classes due to privileged comm. methods

* Now the objection may be raised that these examples don't entail a real change in space. This is because not all modes of communication are affected by the proposed changes. But what could be said if changes were made that affected <u>all</u> modes of communication?

* Here is another example: In 3-D space let us suppose that communication is limited to the horozontal directions. This, in effect, divides 3-D space into a stack of mutually non-communicating 2-D planes.

Now suppose that a class of 2-D beings lives in one of these planes. Also posit a class of 3-D beings for whom communication in the vertical direction is possible.

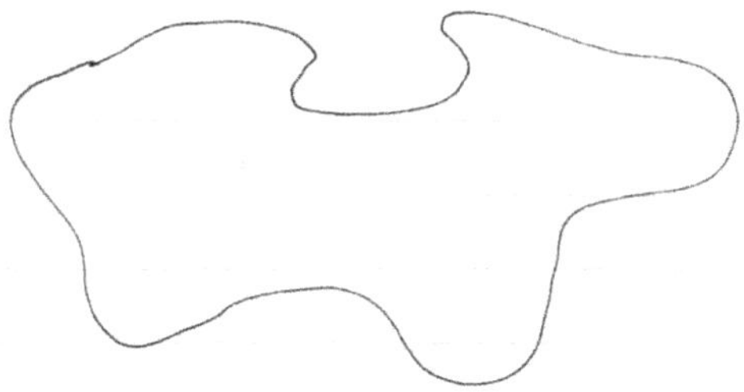

Now suppose that efficient air travel from north to south has become available, but that travel from east to west is only possible by land. There, in effect, the map might be said to have shrunk in the vertical dimension:

- ⋆ Now the objection may be raised that these examples don't entail a real change in space. This is because not all modes of communication are affected by the proposed changes. But what could be said if changes were made that affected all modes of communication?
- ⋆ Here is another example: In 3-D space let us suppose that communication is limited to the horizontal directions. This, in effect, divides 3-D space into a stack of mutually non-communicating 2-D planes.

Now suppose that a class of 2-D beings lives in one of these planes. Also posit a class of 3-D beings for whom communication in the vertical direction is possible.

[Side note] Space different for *devas* due to privileged communication methods]

This is the setting of the flatland story:

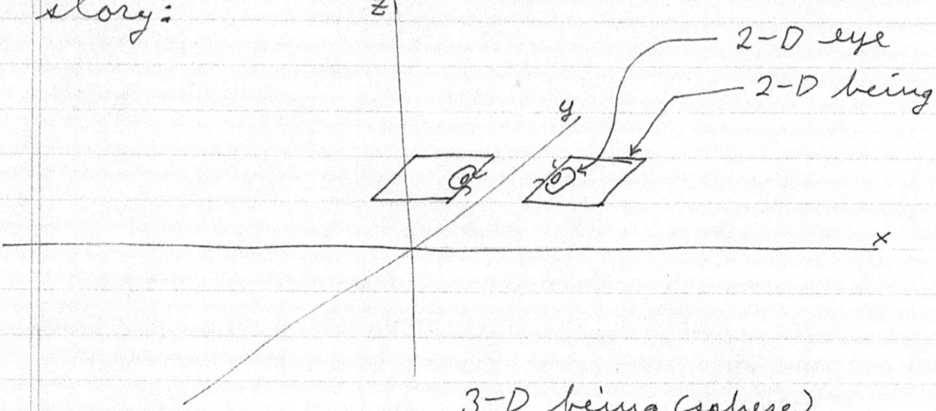

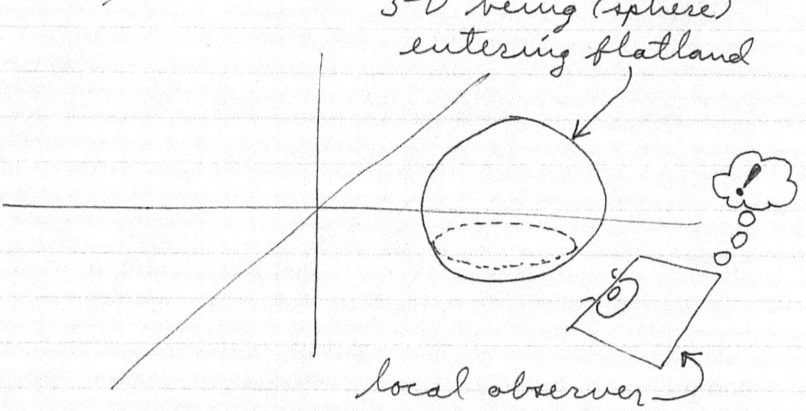

Rod Vrindavan lotus story

Flatland prisoner escaping from prison using "space warp":

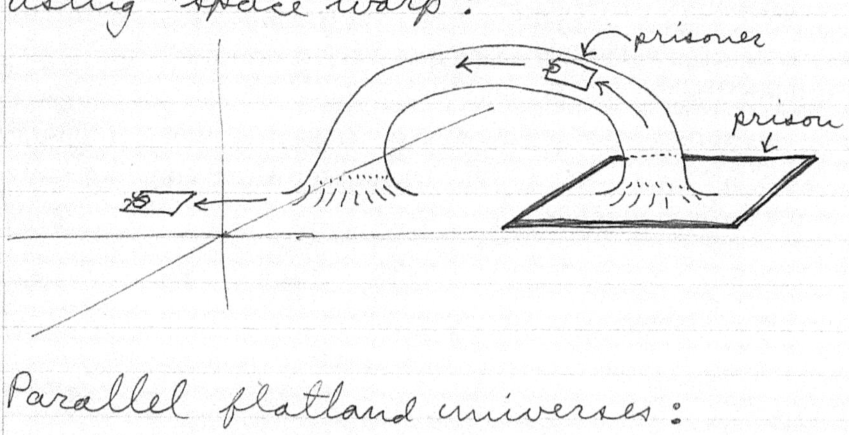

Parallel flatland universes:

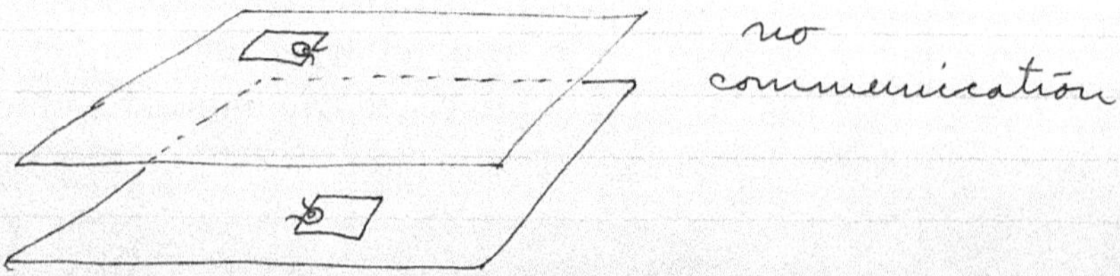

This is the setting of the flatland story:

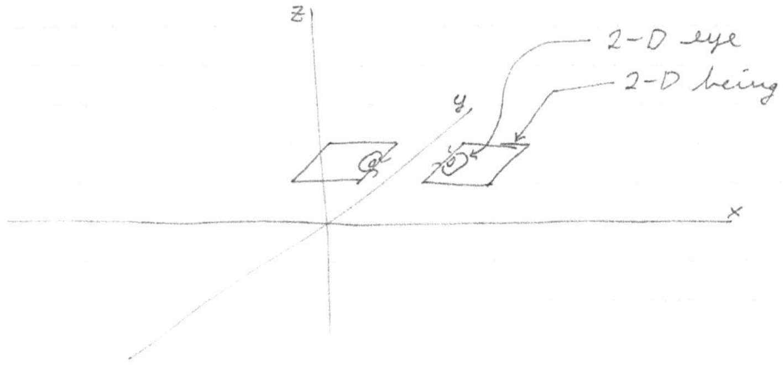

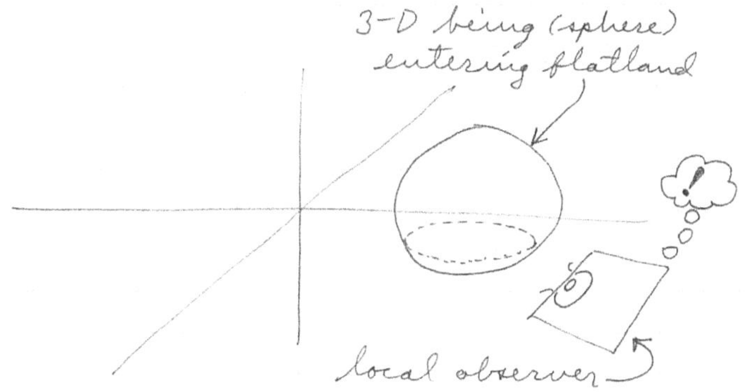

Flatland prisoner escaping from prison using "space warp":

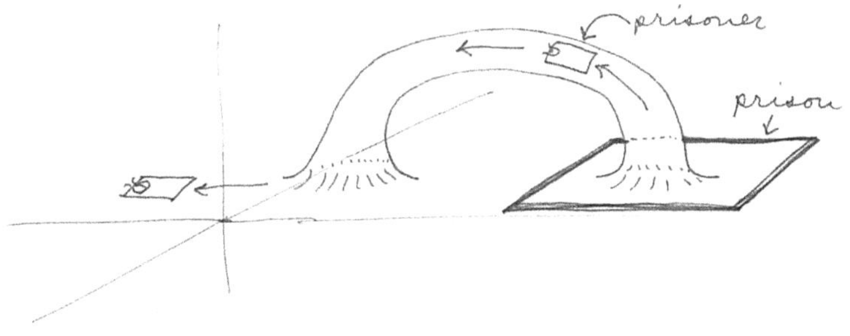

Parallel flatland universes:

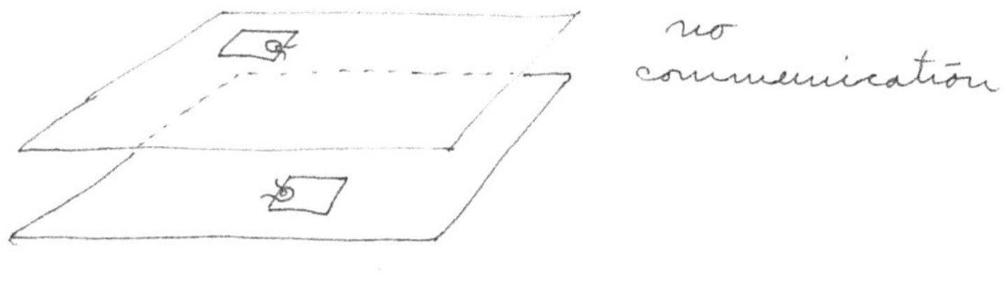

[Side note] Rsd [Ravindra Svarūpa Dāsa] Vṛndāvana lotus story

Parallel flatland universes with connection:

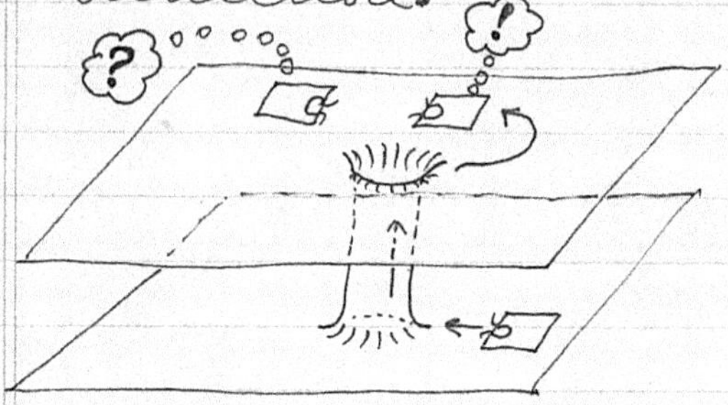

Another flatland parable:

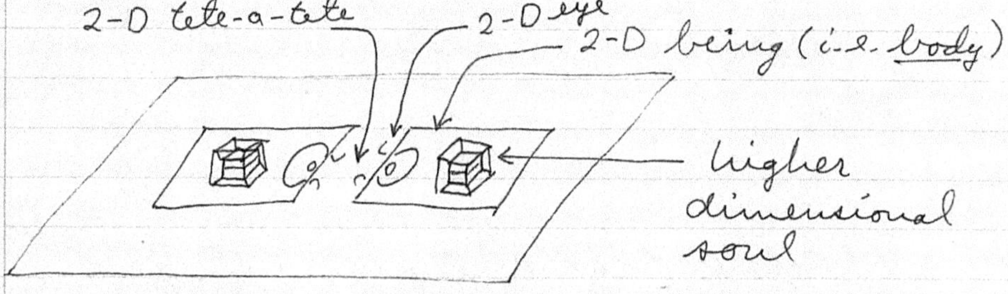

Death in flatland:

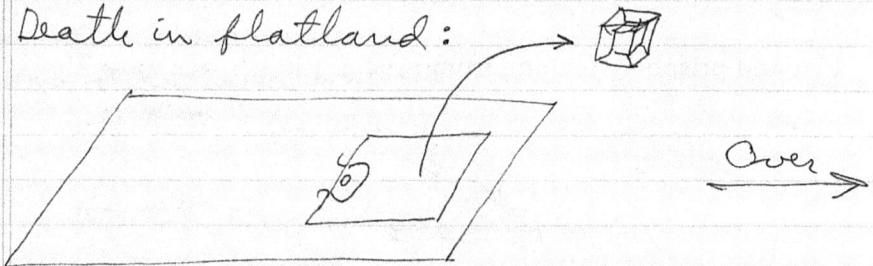

Over →

* Final point on space: In quantum mechanics the states of matter are represented by points in infinite dimensional Hilbert space. Also, the numbers of dimensions needed for a quantum mechanical representation of specific objects is as follows:

1 H atom	3-D
1 He atom	6-D
1 C atom	18-D
1 H_2 molecule	6-D

Parallel flatland universes with connection:

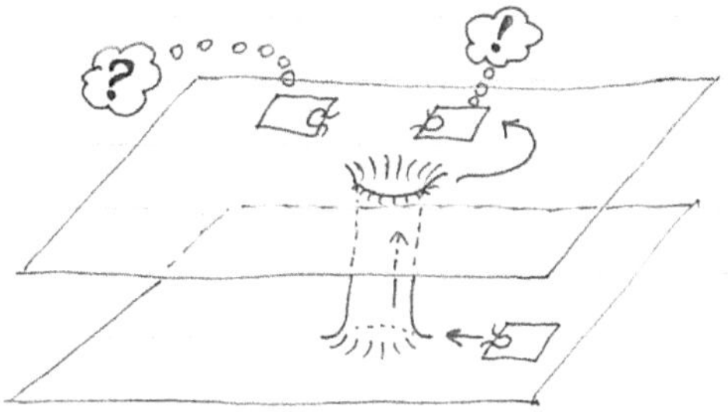

Another flatland parable:

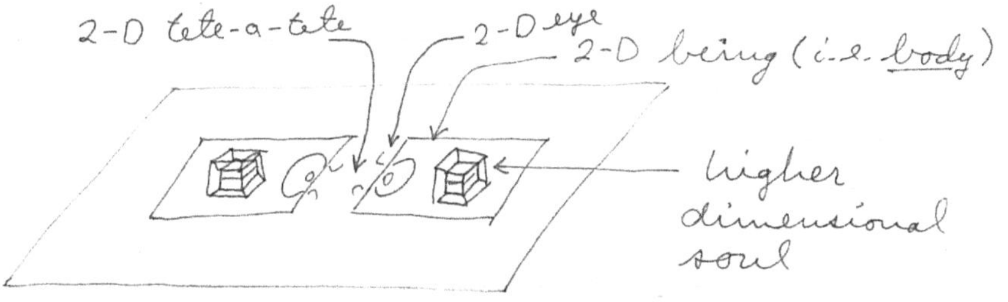

Death in flatland:

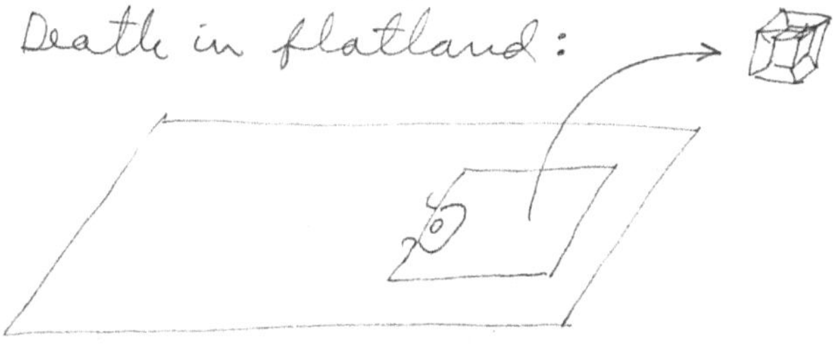

Over →

Another flatland possibility:

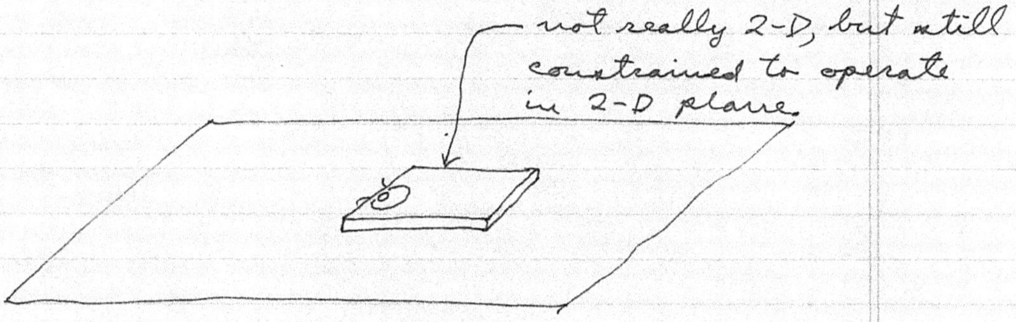

Another flatland possibility:

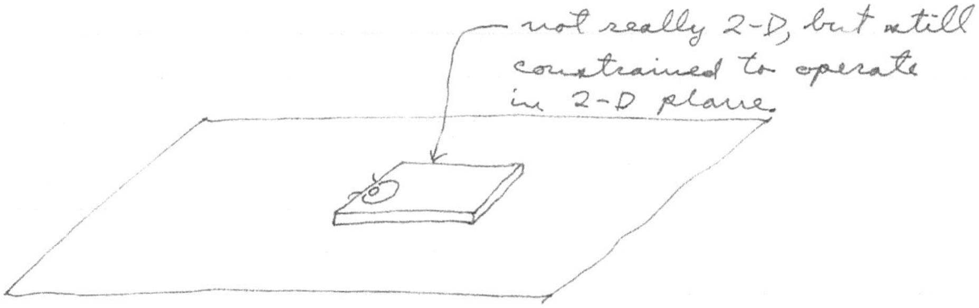

★ Final point on space: In quantum mechanics the states of matter are represented by points in infinite dimensional Hilbert space. Also, the numbers of dimensions needed for a quantum mechanical representation of specific objects is as follows:

1 H atom	3-D
1 He atom	6-D
1 C atom	18-D
1 H_2 molecule	6-D
1 CH_4 molecule	30-D
1 mole of H_2 gas	3×10^{26}-D (approx.)

1 CH_4 molecule 30-D
1 mole of H_2 gas 3×10^{26}-D (approx)

(Here we consider only electrons and we disregard nuclei.)

Of course quantum mechanics is highly abstract. But what kind of reality is it describing?

* Also, note the idea of quantum mechanical tunneling. This is said to occur all the time at the subatomic level. On the macroscopic level, a 1 kg wt. may "tunnel" 1 meter against a 1 newton force (supplied by a spring following Hookes law) with a probability of approx. $10^{-10^{33}}$.

This probability is vanishingly small, but the <u>possibility</u> is interesting. That such "tunneling" is possible is due to the multi-dimensional nature of the quantum mechanical description of nature.

poltergeist prapti siddhi

* Comments on Bohm's example illustrating quantum connectedness:

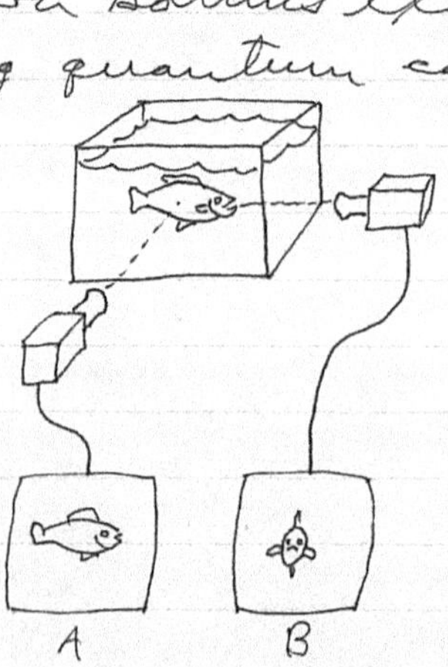

A B

(Here we consider only electrons, and we disregard nuclei.)

Of course, quantum mechanics is highly abstract. But what kind of reality is it describing?

★ Also, note the idea of quantum mechanical tunneling. This is said to occur all the time at the subatomic level. On the macroscopic level, a 1 kg weight. may "tunnel" 1 meter against a 1 newton force (supplied by a spring following Hooke's law) with a probability of approx. 10 to the -10^{33} power.

This probability is vanishingly small, but the possibility is interesting. That such "tunneling" is possible is due to the multi-dimensional nature of the quantum mechanical description of nature.

★ Comments on Bohm's example illustrating quantum connectedness:

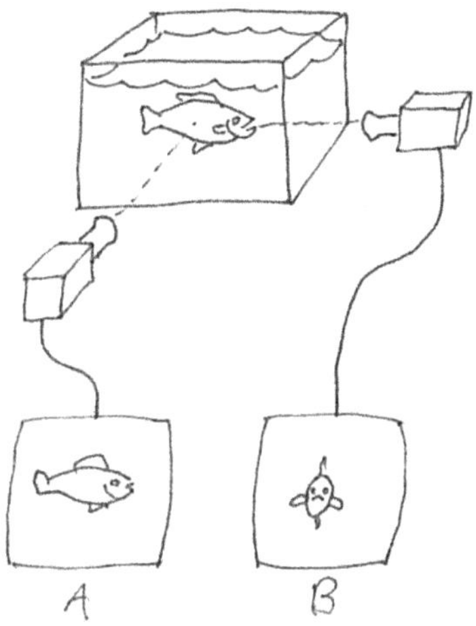

[Side note] poltergeist *prāpti-siddhi*]

(29)

In this example the fish in screens A and B are different *projections* of one actual fish, and thus their activities are strongly coordinated, even though they appear to be different. Note that the real fish is 3-D whereas the fish in screens A and B are 2-D.

(a) Bohm used this idea to illustrate the oneness of the absolute. We note that (1) the original fish has a complex form, and (2) the apparatus of TV cameras and screens is needed to produce the projections. (These are *essential*, not irrelevant, features of the example.)

(b) With this understanding let us generalize the example:

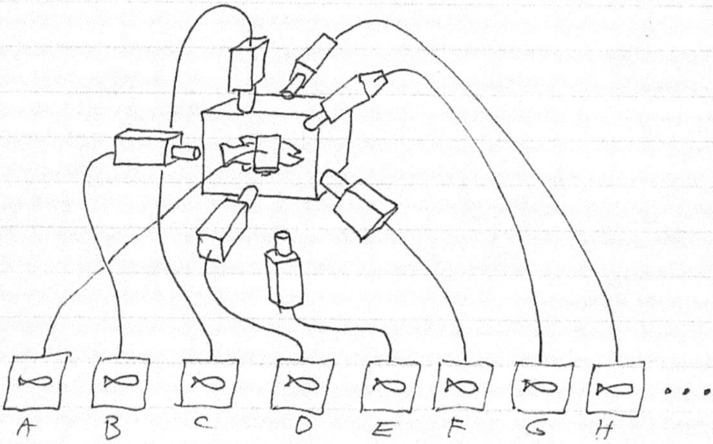

If we introduce many cameras and screens, then most of the views A, B, C, etc of the original 3-D fish will be very similar, for there will be small changes in angle from view to view.

But, if we have n cameras viewing a 3n dimensional fish, then all of the cameras can be situated at right

In this example, the fish in screens A and B are different *projections* of one actual fish, and thus their activities are strongly coordinated, even though they appear to be different. Note that the real fish is 3-D whereas the fish in screens A and B are 2-D.

(a) Bohm used this idea to illustrate the oneness of the absolute. We note that (1) the original fish has a complex form, and (2) the apparatus of TV cameras and screens is needed to produce the projections. (These are essential, not irrelevant, features of the example.)

(b) With this understanding let us generalize the example:

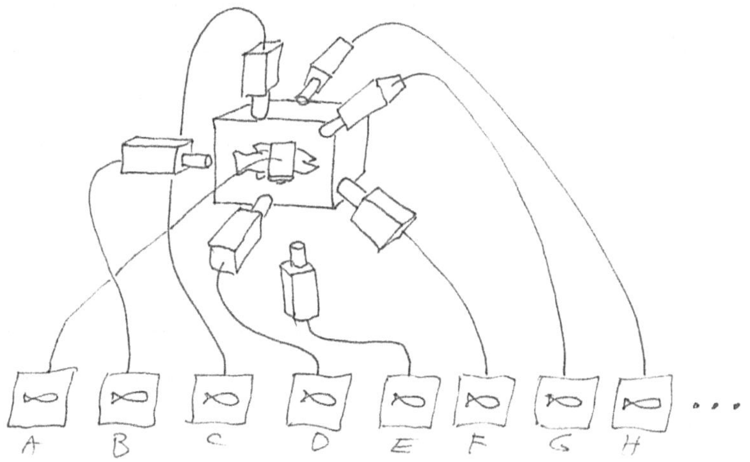

If we introduce many cameras and screens, then most of the views A, B, C, etc. of the original 3-D fish will be very similar, for there will be small changes in angle from view to view.

But, if we have cameras viewing a 3n-dimensional fish, then all of the cameras can be situated at right angles to one another, and all of the projections will be distinct, even though they emanate from one source.

LHQ Thur angles to one another, and all of the projections will be distinct, even though they emanate from one source.

25

* comment on maps: We have argued that natural modes of travel ⇒ natural nbd. relations ⇒ space.
Now, if we give emphasis to a certain mode of travel (communication) then we may wind up with a different conception of the same space.
Example: NYC subway maps:

(a) regular map of Manhattan

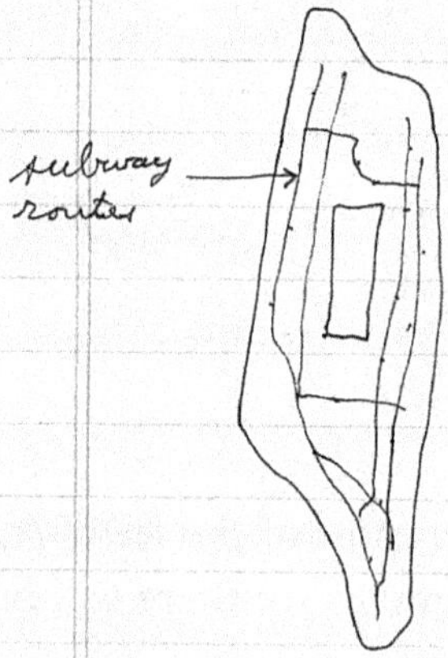

(b) subway map.

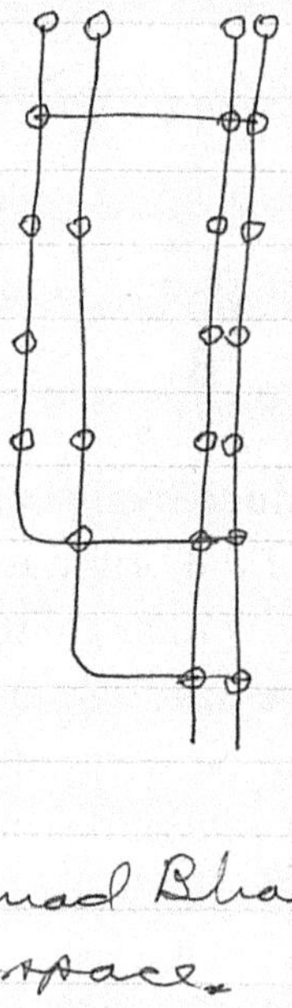

25. Relevance to the Srimad Bhagavatam of these ideas about space.
(a) Yogis have modes of travel that are not available to us, hence they may not see space as we see it. To them we may be comparable to the flatlanders, who are confined to a lower dimensional subspace — both in body and in sensory

★ comment on maps: We have argued that natural modes of travel → natural neighborhood relations → space. Now, if we give emphasis to a certain mode of travel (communication) then we may wind up with a different conception of the same space.

Example: NYC subway maps:

(a)
regular map of
Manhattan

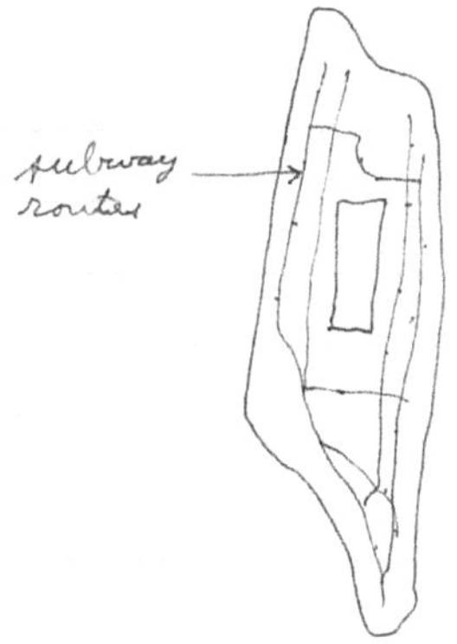

(b)
subway
map.

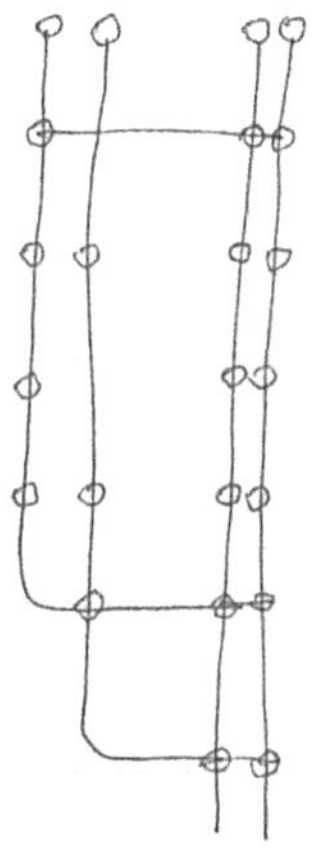

25. Relevance to the *Śrīmad-Bhāgavatam* of these ideas about space.

(a) Yogis have modes of travel that are not available to us, hence they may not see space as we see it. To them we may be comparable to the flatlanders who are confined to a lower dimensional subspace – both in body and in sensory range.

range.

(b) Examples of Vedic evidence for higher dimensionality of space:

- beings with many arms (Banasura) or many heads (Ravana). Note that these are material bodies. They are not even "subtle"—at least not in the case of Banasura who was presumably not firing "subtle" arrows at Kṛṣṇa.

How would this work, anatomically, in three dimensions?

This suggests that space is actually higher dimensional or even trans-dimensional, but that we are normally limited to three dimensions.

(c) Another example: Prapti siddhi. Here, a yogi reaches out his hand, and brings back pomegranits [sic] from Afganistan. Let's say that the Yogi is in NYC at the time. Question: How, exactly, does the fruit travel from Afganistan to NYC? Does it traverse the space in between these points. Does a vastly atenuated hand stretch from NYC to Afganistan, grab the fruits and pull them back through the air?

(b) Examples of Vedic evidence for higher dimensionality of space:

Beings with many arms (Bāṇāsura) or many heads (Rāvaṇa). Note that these are material bodies. They are not even "subtle" – at least not in the case of Bāṇāsura who was presumably not firing "subtle" arrows at Kṛṣṇa.

This suggests that space is actually higher dimensional or even trans-dimensional, but that we are normally limited to three dimensions.

How would this
work, anatomically,
in three dimensions?

(c) Another example: *Prāpti-siddhi.* Here, a yogi reaches out his hand, and brings back pomegranates from Afghanistan. Let's say that the yogi is in NYC at the time.

Question: How, exactly, does the fruit travel from Afghanistan to NYC? Does it traverse the space in between these points? Does a vastly attenuated hand stretch from NYC to Afghanistan, grab the fruits, and pull them back through the air?

(32)

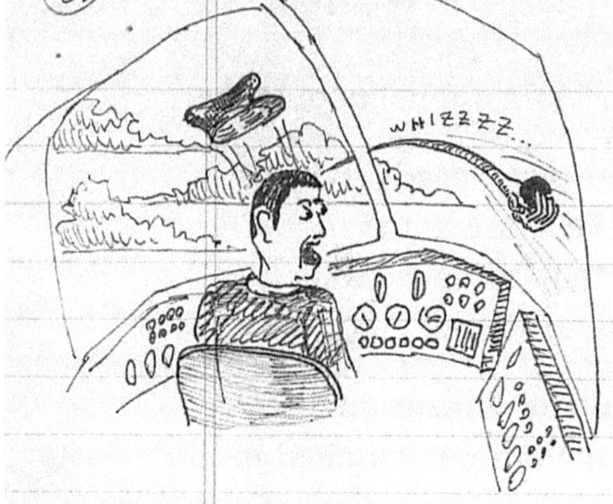

Is an airline pilot likely to see something like this (during a period of peak activity by yogis)?

It would seem that in prapti siddhi the yogis hand remains connected to his body (and the fruit retains its structural integrity) during all stages of the process. Yet the fruit does not traverse the ordinary space between NYC and Afganistan. Therefore, it must traverse some other space (or kind of space).

Vide: the flatland prisoner example. Here is another example — the Ravindra Svarupa Vrindavana story: 'Tis said that the six Gosvamis could traverse distances in Vrindavan in one day on foot which would be too great for us to cover on foot in that time. How did they do it? Well, 'twas said that Vrindavana is like a lotus, and that one can pass from one petal to another if the petals are briefly brought together...

This seems to entail higher dimensions and it can be illustrated by another flatland story:

Is an airline pilot likely to see something like this
(during a period of peak activity by yogis)?

It would seem that in *prāpti-siddhi*, the yogi's hand remains connected to his body (and the fruit retains its structural integrity) during all stages of the process. Yet the fruit does not traverse the ordinary space between NYC and Afghanistan. Therefore, it must traverse some other space (or kind of space).

Vide: the flatland prisoner example. Here is another example – the Ravindra Svarūpa Vṛndāvana story: 'Tis said that the six Gosvāmīs could traverse distances in Vṛndāvana in one day on foot which would be too great for us to cover on foot in that time. How did they do it? Well, 'twas said that Vṛndāvana is like a lotus, and that one can pass from one petal to another if the petals are briefly brought together...

This seems to entail higher dimensions, and it can be illustrated by another flatland story:

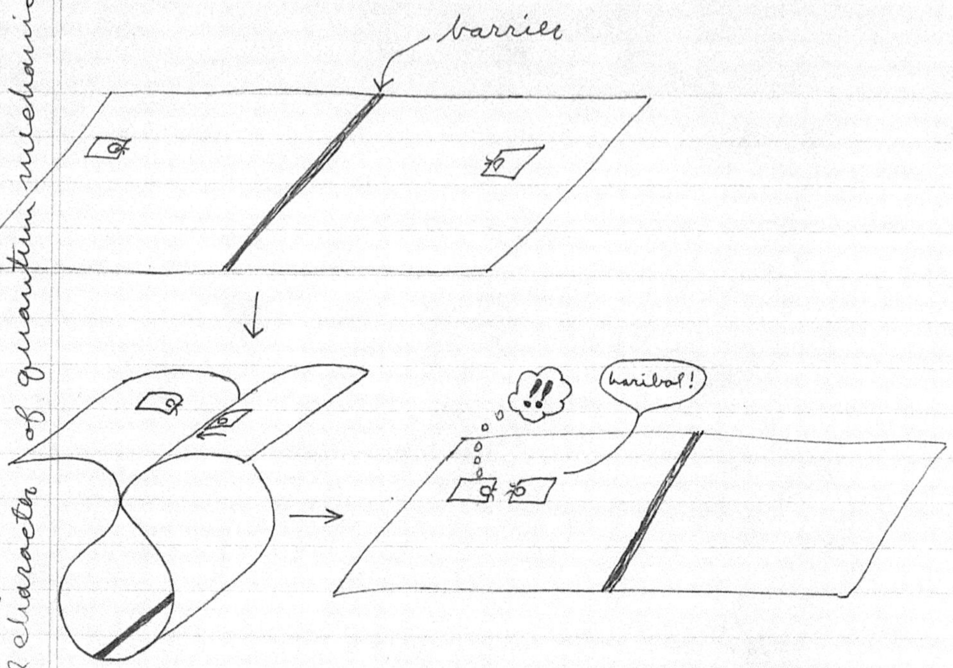

Referring again to prapti-siddhi: The poltergeist phenomena, in which, for example, stones from the garden mysteriously fall from the ceiling, seem to entail something similar to prapti-siddhi. Also, it is said that certain "yogis" such as Satya Sai Baba, have mastered the prapti-siddhi and use it to prove to people that they are "God." Thus, the existence of the prapti-siddhi is supported by both Vedic evidence and current sensory evidence.

Note also: Quantum mechanical tunneling of large objects, although of immensely small probability, is a kind of prapti-siddhi that could exist in principle, and that depends on the higher dimensional character of quantum mechanics.

(d) Another example: The story of Kalapa grama in the S.B. is an example of references to "parallel worlds" in the Vedic literature. The Navadvipa-mahatmya provides an even more striking example. Vide the flatland "parallel universe" example.

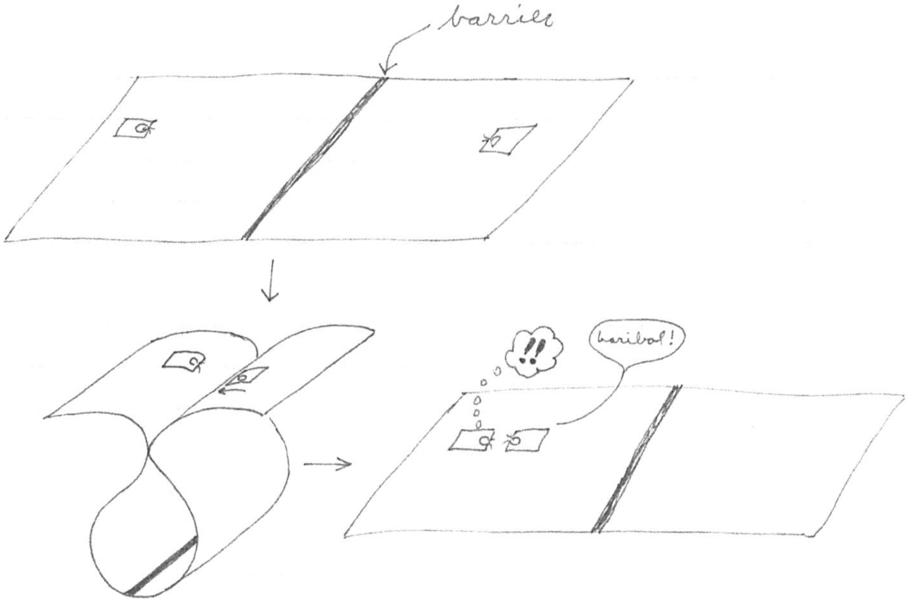

Referring again to *prāpti-siddhi*: The poltergeist phenomena, in which, for example, stones from the garden mysteriously fall from the ceiling, seem to entail something similar to *prāpti-siddhi*. Also, it is said that certain "yogis" such as Sathya Sai Baba, have mastered the *prāpti-siddhi* and use it to prove to people that they are "God." Thus, the existence of the *prāpti-siddhi* is supported by both Vedic evidence and current sensory evidence.

(d) Another example: The story of Kalāpa-grāma in the *S.B.* is an example of reference to "parallel worlds" in the Vedic literature. The *Navadvīpa-dhāma-mahātmya* provides an even more striking example. Vide: the flatland "parallel universe" example.

[Side note] Note also: Quantum mechanical tunneling of large objects, although of immensely small probability, is a kind of *prāpti-siddhi* that could exist in principle and that depends on the higher dimensional character of quantum mechanics.]

N.B. Quantum mechanical parallel universe theory.

(e) Application to the 5th Canto:
How do people travel from place to place in the 5th Canto universe? How, for example, did the people from the seven dvipas who visited Lord Caitanya as an infant, travel to Mayapur? It seems unlikely that they traveled by ship across the salt ocean and docked in the port of Calcutta. It would seem that they traveled by a more direct method that transcends our ideas of space...

Also, how does Narada Muni travel from place to place?

26. We suggest that the 5th Canto universe is a higher dimensional or transdimensional realm, and that it is represented spatially as it would appear to a person who can travel by yogic siddhis. The 5th Canto description may even be somewhat stylized, as in a NYC subway map, which is oriented exclusively to subway travel.

The relative distances to the sun, moon, etc. are depicted as they would appear to traveling yogis. Note that the Vedanta-sutra commentary of Baladeva Vidyabhusana describes how the transmigrating soul passes through a series of destinations, including the sun and then the moon, on its way to the higher planets. Here we clearly have a mode of

(e) Application to the 5th Canto: How do people travel from place to place in the 5th Canto universe? How, for example, did the people from the seven *dvīpas* who visited Lord Caitanya as an infant, travel to Māyāpura? It seems unlikely that they traveled by ship across the salt ocean and docked in the port of Calcutta. It would seem that they traveled by a more direct method that transcends our ideas of space...

Also, how does Nārada Muni travel from place to place?

26. We suggest that the 5th Canto universe is a higher dimensional or trans-dimensional realm, and that it is represented spatially as it would appear to a person who can travel by yogic *siddhis*. The 5th Canto description may even be somewhat stylized, as in a NYC subway map, which is oriented exclusively to subway travel.

The relative distances to the sun, moon, etc., are depicted as they would appear to traveling yogis. Note that the *Vedānta-sūtra* commentary of Baladeva Vidyābhūṣaṇa describes how the transmigrating soul passes through a series of destinations, including the sun and then the moon, on its way to the higher planets. Here we clearly have a mode of travel which is beyond our ordinary experience.

[Side note] N.B. Quantum mechanical parallel universe theory travel which is beyond our ordinary experience.

The Surya siddhanta, in contrast, describes the solar system as it appears to a human being with ordinary sensory endowments.

27. More evidence pertaining to the nature of "Vedic space": Srila Prabhupada has said that a yogi can enter the Ganges in the Himalayas, and emerge at the Magh mela in Allahabad. Clearly he doesn't do this by swimming downstream under water. A yogi can also travel to the celestial planets via the Ganges.

This also entails a higher mode of travel, and hence a higher form of space.

# of heads on Brahmas...

28. Note that space is <u>created</u>, according to the S.B. It is not absolute. (Space is, in fact, akasa or the element, ether.) Hence the spiritual world must completely transcend material space. [If this is so, then why can it not be that material space, as a whole, contains features not accessible to beings with our sensory and bodily equipment? That is, might it not be that our physical limitations restrict us to a subspace of the total material space, similar to the subspace of the flatlanders?]

The *Sūrya-siddhānta*, in contrast, describes the solar system as it appears to a human being with ordinary sensory endowments.

27. More evidence pertaining to the nature of "Vedic space": Śrīla Prabhupāda has said that a yogi can enter the Ganges in the Himalayas, and emerge at the *Māgh-melā* in Allahabad. Clearly he doesn't do this by swimming downstream under water. A yogi can also travel to the celestial planets via the Ganges.

This also entails a higher mode of travel, and hence a higher form of space.

28. Note that space is *created*, according to the *S.B.* It is not absolute. (Space is, in fact, ākāśa or the element, ether.) Hence the spiritual world must completely transcend material space. If this is so, then why can it not be that material space, as a whole, contains features not accessible to beings with our sensory and bodily equipment? That is, might it not be that our physical limitations restrict us to a subspace of the total material space, similar to the subspace of the flatlanders?

[Side note] # of heads on Brahmās

29. Evidence pertaining to the nature of transcendental space [which must include material space, since the mahat tattva occupies a corner of the spiritual sky.]:

(a) The story of Kṛṣṇa meeting the Brahmas in Dwaraka. Here one room in Dwaraka was present in many universes simultaneously.

(b) The Brahma mohana lila.

(c) The manifestation of the universal form in *Bhagavad-gita*.

(d) The appearance of all the universes in Kṛṣṇa's mouth, when mother Yasoda checked to see if He was eating dirt. Note that Mother Yasoda saw herself and Kṛṣṇa within Kṛṣṇa's mouth.

(e) The verses of the *Brahma-samhita*, such as
eko'py asau racayitum
jagad-anda-kotim ...

(f) Kṛṣṇa's multiple expansions in many palaces in Dwaraka (and let us also remember the bodyguards of Maharaja Ugrasena in Dwaraka...)

(g) The principle of acintya bhedābheda tattva.

yogis must enter "atmosphere" of planets...

29. Evidence pertaining to the nature of transcendental space, which must include material space, since the *mahat-tattva* occupies a corner of the spiritual sky:

(a) The story of Kṛṣṇa meeting the Brahmās in Dvārakā. Here one room in Dvārakā was present in many universes simultaneously.

(b) The *brahma-vimohana-līlā*.

(c) The manifestation of the universal form in *Bhagavad-gītā*.

(d) The appearance of all the universes in Kṛṣṇa's mouth, when mother Yaśodā checked to see if He was eating dirt. Note that Mother Yaśodā saw herself and Kṛṣṇa within Kṛṣṇa's mouth.

(e) The verses of the *Brahma-saṁhitā,* such as *eko 'py asau racayituṁ jagad-aṇḍa-koṭiṁ*... [Bs. 5.35]

(f) Kṛṣṇa's multiple expansions in many palaces in Dvārakā (and let us also remember the bodyguards of Mahārāja Ugrasena in Dvārakā...)

(g) The principle of *acintya-bhedābheda-tattva*.

[Side note] yogis must enter "atmosphere" of planet ...

(37)

30. So, if material space is a "subspace" of trans-dimensional transcendental space, then why should we expect that it can be fully described by 3-D Euclidean geometry?

31. Let us refer back to the Bohm-fish example. According to the *Bhagavad-gita*, the supersoul appears to be divided into many, but actually he is situated as one. This example, which makes use of the idea of multidimensionality, may have its real application here (imperfect though it must be.)

32. Note that the jiva-soul is trans-dimensional, but is somehow associated with the gross body. Think about this in the context of the flatland parables.

30. So, if material space is a "subspace" of trans-dimensional transcendental space, then why should we expect that it can be fully described by 3-D Euclidean geometry?

31. Let us refer back to the Bohm-fish example. According to the *Bhagavad-gītā*, the Supersoul appears to be divided into many, but actually he is situated as one. This example, which makes use of the idea of multi-dimensionality, may have its real application here (imperfect though it must be).

32. Note that the *jīva*-soul is trans-dimensional, but is somehow associated with the gross body. Think about this in the context of the flatland parables.

Other questions and observations concerning astronomy:

1. The multi-dimensional space theory that we are considering may also be useful for understanding the nature of: ghosts, "cryptozoans" (elusive animals such as "bigfeet", etc.), UFO's etc. Of course, all of these items are highly disreputable in orthodox circles.

2. The 5th Canto view of the universe receives support from folklore and "mythology". For example:
   - The Jack & the beanstalk story features Rakṣasas...
   - Norse cosmology gives the following picture of the earth:

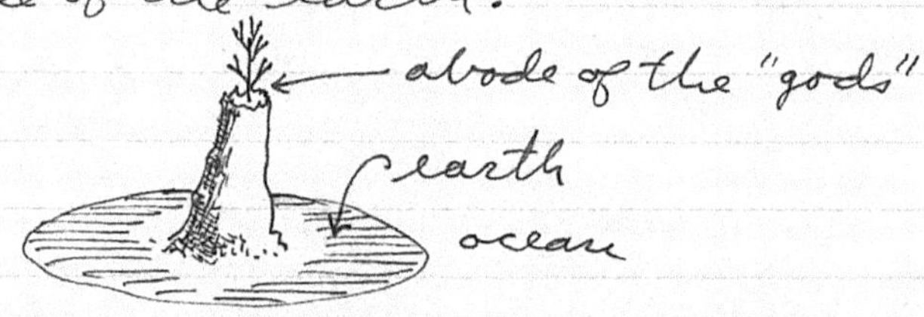

   - Norse elves correspond to Gandharvas
   - Mt Olympus of the Greeks corresponds to Mt Meru
   - The Mayan Indian cosmology is similar to that of the 5th Canto. (Also, the Mayan calendar begins in ≈ 3113 B.C. Compare this with the starting date of the Kaliyuga in 3102 B.C.)

Other questions and observations concerning astronomy:

1. The multi-dimensional space theory that we are considering may also be useful for understanding the nature of: ghosts, "cryptozoans" (elusive animals such as "bigfoot," etc.), UFOs, etc. Of course, all of these items are highly disreputable in orthodox circles.

2. The 5th Canto view of the universe receives support from folklore and "mythology." For example:

   ⋆ The "Jack and the Beanstalk" story features Rākṣasas...

   ⋆ Norse cosmology gives the following picture of the earth:

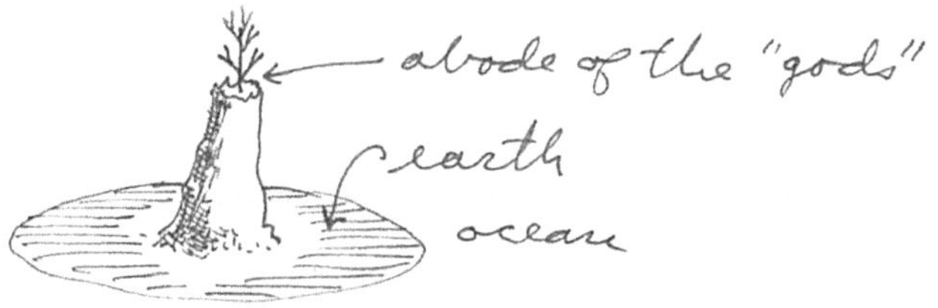

   ⋆ Norse elves correspond to Gandharvas.

   ⋆ Mt. Olympus of the Greeks corresponds to Mt. Meru.

   ⋆ The Mayan Indian cosmology is similar to that of the 5th Canto. (Also, the Mayan calendar begins in ~3113 B.C. Compare this with the starting date of the Kali-yuga in 3102 B.C.)

(39)

3. Srila Prabhupada has stressed that no animals have become extinct. Where are the apparently extinct animals then? According to our interpretation of the 5th Canto, there are many inhabited lands which are not accessible to us, and many life forms could be living there. To illustrate this idea, note that wolves have become extinct in, say, Long Island, but they are still living in the Rocky Mtns.

4. The description of Brahma's assembly house in the Mahabharata strongly conveys the impression of multidimensionality, or trans-dimensionality. This is true, as well, of many other stories in the Mahabharata (which can be studied as a source of ideas in "higher physics".)

5. Here we consider the following problem: The S.S. is geocentric, whereas modern astronomy is heliocentric (as far as the solar system is concerned.) If we adopt the S.S. as an observational description of the solar system, then what do we do about this. Here I will not answer this question. I will only make the following observations:

(a) The Vedas certainly seem to regard the earth as fixed and the sun as moving.

3. Śrīla Prabhupāda has stressed that no animals have become extinct. Where are the apparently extinct animals then? According to our interpretation of the 5th Canto, there are many inhabited lands which are not accessible to us, and many life forms could be living there.

   To illustrate this idea, note that wolves have become extinct in, say, Long Island, but they are still living in the Rocky Mountains.

4. The description of Brahmā's assembly house in the *Mahābhārata* strongly conveys the impression of multi-dimensionality, or trans-dimensionality. This is true, as well, of many other stories in the *Mahābhārata*, which can be studied as a source of ideas in "higher physics."

5. Here we consider the following problem: The *S.S.* is geocentric, whereas modern astronomy is heliocentric (as far as the solar system is concerned). If we adopt the *S.S.* as an observational description of the solar system, then what do we do about this. Here I will not answer this question. I will only make the following observations:

   (a) The *Vedas* certainly seem to regard the earth as fixed and the sun as moving.

Yet: (b) The formulas used in the S.S. are essentially heliocentric. Here is why: The orbits of the sun and moon are roughly as follows (in these diagrams we disregard what is known in the S.S. as the mandoccha.):

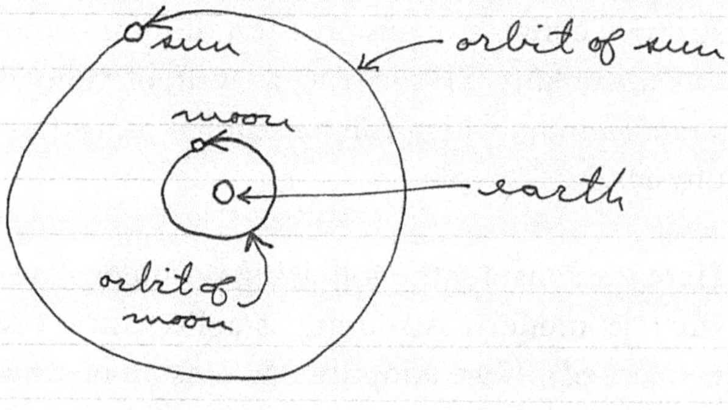

The orbits of mercury and venus are:

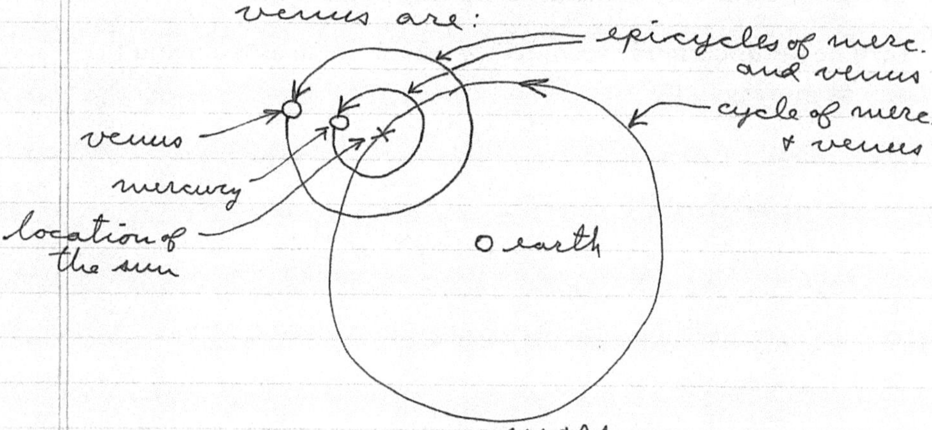

The orbits of mars, jupiter and saturn are

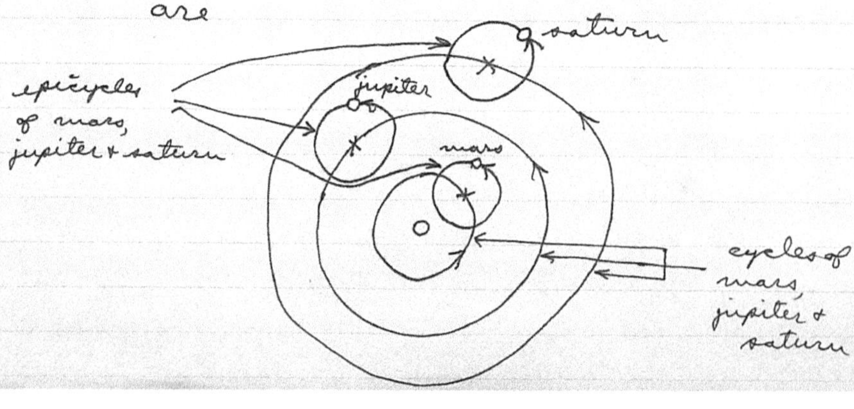

Yet:

(b) The formulas used in the *S.S.* are essentially heliocentric. Here is why: The orbits of the sun and moon are roughly as follows (in these diagrams we disregard what is known in the *S.S.* as the *mandoccha*):

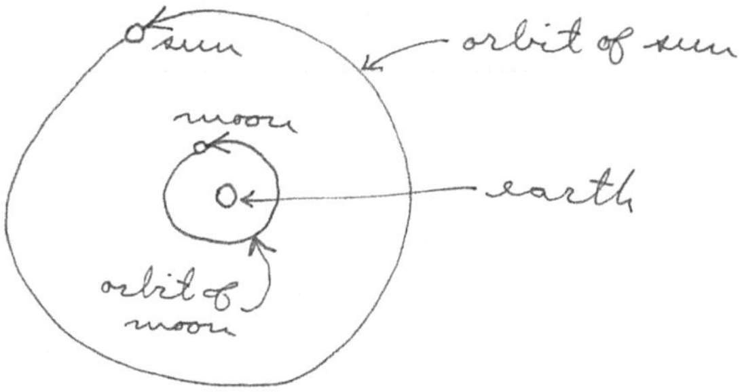

The orbits of Mercury and Venus are:

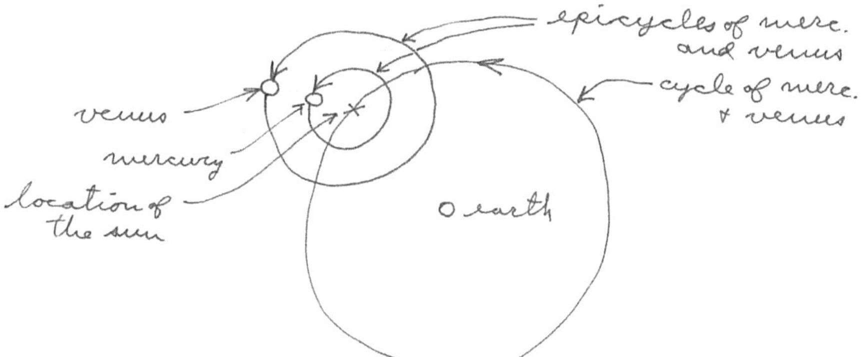

The orbits of Mars, Jupiter and Saturn are:

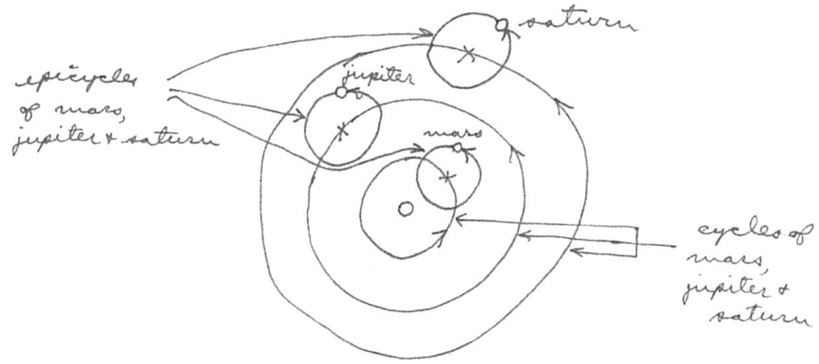

The periods of these cycles and epicycles are related as follows:

planet	period of cycle	period of epicycle
sun	1 earth year	none
moon	1 lunar month	none
mercury	1 year of mercury	1 earth year
venus	1 year of venus	1 earth year
mars	1 earth year	1 year of mars
jupiter	1 earth year	1 year of jupiter
saturn	1 earth year	1 year of saturn

Why are the cycle periods of mars jupiter and saturn all = 1 earth year?
Why are the epicycle periods of mercury and venus = 1 earth year?
The heliocentric model provides a simple explanation, whereas the geocentric model provides no explanation. This was one of Copernicus's reasons for accepting the heliocentric model.
Note: Aryabhatta (6th cent. A.D.) presented a heliocentric version of the Surya-siddhanta.
Note also: The navy still uses a geocentric solar system model for navigational purposes.

The periods of these cycles and epicycles are related as follows:

Planet	Period of cycle	Period of epicycle
Sun	1 earth year	none
Moon	1 lunar month	none
Mercury	1 year of Mercury	1 earth year
Venus	1 year of Venus	1 earth year
Mars	1 earth year	1 year of Mars
Jupiter	1 earth year	1 year of Jupiter
Saturn	1 earth year	1 year of Saturn

Why are the cycle periods of Mars, Jupiter and Saturn all = 1 earth year?

Why are the epicycle periods of Mercury and Venus = 1 earth year?

The heliocentric model provides a simple explanation, whereas the geocentric model provides no explanation. This was one of Copernicus' reasons for accepting the heliocentric model.

Note: Āryabhaṭa (6th century A.D.) presented a heliocentric version of the *Sūrya-siddhānta*.

Note also: The navy still uses a geocentric solar system model for navigational purposes.

CHAPTER 9

On the Relationship Between Consciousness and Matter

Presented at a meeting of science faculty at
Eindhoven University of Technology
Eindhoven, Netherlands / August 8, 1985

Today I am just planning to say a few words about the relationship between consciousness and matter, and I want to give a report on some investigations that I have done recently in connection with the quantum theory, and show how this ties in with Vedic philosophy. I will begin by saying a few words about consciousness. This actually is an illustration from my book, if you want to see the projection here (MNS, Chapter 1, Fig. 1).

One may ask: What is the nature of the relationship between states of consciousness and material systems? In modern science, of course, the idea has been very widely disseminated that consciousness is a product of the action of the brain.

Generally, people think in terms of the computer model in which various interactions involving nerve impulses and so forth generate the functions of the mind. However, it has also been observed that there is such a thing as consciousness, in and of itself. This is a whole subject of discussion of philosophy, known basically as the mind-body problem. Of course many volumes have been written on this subject.

I am not going to try and review the discussions of the mind-body problem here, but I wanted to make a brief observation concerning the kind of relationship that must exist between consciousness and physical states of the brain, if we accept consciousness as an actually existing reality. The basic point here is that in analyzing the brain, one can look at it in terms of a hierarchy of levels of abstraction. At the lowest level, as shown in this chart, you have the brain hardware – this is indicated here by some drawings of nerve cells (MNS, Chapter 1, Figures 2 and 3).

Within this brain hardware, elementary physical events occur which supposedly should add up to generate all the different phenomena that are produced by the brain. For example, nerve impulses can travel along different axons and dendrites of the neurons and synapses, and different chemical reactions can take place, and so on. So, these are basic neural operations.

To try and understand how the brain could represent the mind, the general approach that is used is to organize conceptually basic operations into larger units, which can be referred to by symbols in a higher order language of some kind. Now of course, this has not been really done yet for the brain, but this is the approach people use when they deal with computers.

For example, I will just share one illustration here from a book on artificial intelligence. This is probably quite small, but in the series of levels here – the lowest one says transistors, as in an old fashioned transistorized computer. Then from the level of flip flops and gates – now flip flops and gates are actually combinations of transistors – so that represents a level of abstraction above the transistor level. Then you come to registers and data pads – machine instructions – this is the level of machine language, a compiler or interpreter. And then LISP – that is a kind of programming language.

So at the level, say of LISP, you will have statements in a language, but they refer to complicated combinations of the lower order entities in this hierarchy. Ultimately, they refer to actual physical processes in a machine. And finally, according to this artificial intelligence scheme, you get up to intelligent programs at the top.

Of course this has not really been realized yet, but that is the idea of how one could represent mind in terms of the operation of a machine. So going back to this picture, let us suppose you can do this for the brain. Then at a lower level you would have basic neural operations. You can work your way up some kind of hierarchy until at the upper level you would have things corresponding to thoughts, feelings, perceptions, and so forth. (MNS, Chapter 2, Fig. 4)

If we regard consciousness as being actually something that really exists, that is, we have states of consciousness – for example I am now perceiving people here in this room and also perceiving what I am feeling and so on – so the question comes: What is the relationship between consciousness and the material system? One can say that whatever the nature of that relationship is, it can only be a direct relationship between the states of consciousness – thoughts, feelings, and so on – and the higher order patterns within this material system. There is no direct one-to-one correlation between the states of consciousness and, say, nerve impulses.

This leads to the idea that the relation between the states of consciousness and the material system is in some way what the physicists would call nonlocal. I will mention briefly what this concept is. In physics there is the idea that material entities can either interact with one another in their immediate vicinity – that is with no spatial separation – or they might interact with one another across a distance.

Of course, when Newton introduced his law of gravitation, it was very controversial because he was proposing that gravitational force acts instantly over a large distance. So that is an example of a nonlocal interaction. An example of a local interaction would be given by, say, the laws of electromagnetism. If you have an electric and magnetic field, then according to Maxwell's equations, at each point they are interacting with each other with no spatial separation.

So in present day physics, the basic idea is that all interactions should be understood as local interactions. This is actually required by Einstein's special theory of relativity, for example, and so physicists are very uncomfortable about nonlocal interactions. In particular, of course, a local theory of gravitation was introduced by Einstein; that is his general theory of relativity.

The point I wanted to introduce here about consciousness is that if you accept consciousness as something real, it seems to interact with matter in a nonlocal fashion. So, I am going to return to that point later on in the discussion.

Now before going further with consciousness, I want to introduce another basic theme which I would like to discuss, and that is the theme of biological complexity. So here we have ... [*break in recording*] ... mechanical arrangements within the bodies. For example, the eye has many highly complicated systems. I was told by one medical student that I knew that there is a certain membrane in here called Descemet's membrane, just one little layer of cells, and somebody wrote a three hundred page doctoral dissertation just on that.

These biological structures can be highly complex, and the question is: How do such structures originate? So of course, this has been a subject of much discussion. And the basic idea of how they originate, which has been adopted by modern day science, is the Darwinian theory of evolution and various forms of that. It is a fact, however, that this theory has not provided explicit explanations of how highly complex forms come into being.

Actually, all of the examples in which one can give a plausible explanation of how biological form comes into being through an evolutionary process involve very simple aspects of biological form. Of course, one can say that a complex form is difficult to deal with simply because it is complex, and indeed that is true. But it remains a fact, that it is somewhat of a mystery at the present time, how these complex forms come into being.

This is another interesting example here: It seems that there are certain butterflies which have iridescent wings. If you magnify the wings, you will see they are made of little scales like shingles of a roof. If you magnify them further, you will find that there are very tiny diffraction gratings mounted on raised platforms along the lengths of these scales. So somehow or the other, nature has created diffraction gratings there. One may wonder just how this comes about.

Some indication of what is involved in the origin of biological form, is given by the study of biological form on the molecular level. For example, this is a structural diagram of a transfer RNA molecule – it is one of the molecules that exist within living cells. It is a very complex and precisely constructed molecule. And in fact, on the molecular level within cells, one finds a very high level of complexity.

Here is an example of a bacteriophage. This is a simple diagram of a kind of virus that infects bacteria – and I can show briefly here how the bacteriophage works. This is interesting because it is an example of a fairly simple biological system, in fact, one of the simplest in existence. This device consists of a hollow shell containing the DNA of the bacteriophage, and a kind of syringe mechanism.

What happens here is that the bacteriophage will come in contact with a wall of a bacterium, and it recognizes the surface as being that of the right kind of bacterium. Then a spring-like arrangement contracts and drives this hollow syringe through the wall of the bacterium; and the DNA, which is coiled up inside this capsule, goes through the tube into the bacterium. Once that happens, the bacterium begins to follow the instructions in the viral DNA, and it begins to manufacture viruses through a process of assembly of different parts.

Just for the fun of it, here is an example of a very simple simulation of such a process; this is a very super-simplified bacteriophage. We have rectangles with bond sites, and these represent molecules – protein molecules – which can bond together according to certain rules. These are so designed that once they are allowed to move about at random, they can actually bond together to form a completed structure like this.

The process of this bonding together is shown here in a computer simulation. This is a time sequence in which you start out with a random arrangement of these molecules, and gradually they begin to assemble together to form a completed structure by engaging in a random walk process. So this is a simulation of this on a very simple level, of what is going on with this bacteriophage.

But the interesting point to make about this is that you can ask: What happens if you vary the parameters which are involved in specifying these different bond values, and the different distances, and so forth?

What you find is that there are only certain combinations of the parameters that will produce a system that works and assembles itself. Most combinations do not do so. In fact just varying two of the parameters, we find a situation like this. In this particular graph, two parameters are applied on the horizontal axes here, and the vertical axis represents the degree of effectiveness with which the system functions and assembles together. And one can see that you have a series of peaks that tend to be separated from one another.

This turns out to be a general property of such systems: There will be many different parameters, and only for certain very precisely specified combinations will the system function properly. If you vary them very slightly, then the system ceases to function properly. You can imagine a multi-dimensional space, with many different parameters representing all the different properties of the system. There will be many isolated islands or peaks within that space representing viable systems, and a very large volume of space representing systems that do not work.

So the question is: By what process do you manage to find your way to a peak, or an island, representing a viable form? By an evolutionary optimization process, once you are on a given peak you can climb up to the top moving in the immediate vicinity. But the question is: How do you get to such a peak? This is a general question in biology.

So what I wanted to discuss was some alternative ideas to the standard explanation of how this comes about. One example of such an idea is the theory propounded by a man named Walter Elsasser at Johns Hopkins University. Elsasser was a physicist who later turned to biology. He proposed a rather interesting theory as to how complicated organic form could come into being, based on some of the unique features of the quantum theory.

Briefly, the nature of the quantum theory is that it does not determine exactly what is going to happen in a physical system. You are no doubt familiar with quantum phenomena such as radioactive decay, in which the particular time in which a radioactive atom, say, emits an alpha particle, is not determined by any physical law. It is a completely random event.

Well, random events as treated by quantum mechanics can be dealt with statistically, as long as the probabilities for these events are fairly high. For example, if you have something with a probability of one out of a thousand, then if you repeat the circumstances in which that event can arise several thousand times, and the event happens several times, then you have evidence that it is happening with a probability of about one out of a thousand. If you could only repeat it a hundred times, and say it occurred once, then you would have no basis for saying that it had a probability of one out of a thousand.

So, to talk about probabilities of events, you need to have repetitions commensurate with the reciprocal of the probability, that is, one divided by the probability. So in quantum mechanics it turns out that probabilities, or possibilities, are there for very large numbers of distinct events, so that these have nearly equal probability.

The numbers can be so large that they are what Elsasser called "immense." He used the term "immense" to refer to any number larger than about 10^{100}. If you have a situation in which there are more than 10^{100} alternatives, and one of them comes up, then you cannot say that that is happening by chance, because it is impossible to reproduce the situation enough times to measure such probabilities.

Elsasser used this idea as a means of introducing new laws into physics which are compatible with the laws of quantum mechanics, but which are independent of those laws. Actually, Elsasser proposed that the laws of quantum mechanics, as they stand, are quite correct. But he said, though, that one could have additional laws.

This was his basic idea, and he referred to these laws by the term "creative selection." That is, he said that there can be a process in nature which simply selects highly complex forms. This process does not depend on the laws of physics – it cannot be deduced from them. But also, it does not contradict them. It is an independent aspect of nature, or of reality, to which you can add the existing framework of quantum mechanics.

So now I would like to say a little bit about quantum mechanics. What I am going to do is elaborate a little bit on Elsasser's idea and show how that ties up with other interesting ideas.

First, I will introduce the idea of the quantum mechanical wave function. In quantum mechanics, one does not specify exact positions and velocities of particles; one specifies something called a wave function which is spread out in space. So here we have a particle, and in quantum mechanics it can be spread out. Now this spreading is a rather interesting phenomenon. If it occurs just on an atomic or microscopic level there is no problem, but it is possible for this spreading to occur on a macroscopic level.

This is exemplified by the famous Schrödinger cat paradox. This is perhaps the basic example of this. In this paradox – well, it is called a paradox – you have some radioactive atoms, a Geiger counter, and then some apparatus which will kill a cat – which is within a box – if a radioactive atom decays within a certain time period. And on the other hand, if the atom does not decay within that time period, then the cat is not killed.

So according to quantum mechanics, the state of the radioactive atom is actually indeterminate. It is described by a wave function, like this, which spreads out not over the position exactly, but over the state of the atom being decayed or not decayed. The state of the atom in being partly decayed and partly not decayed, the wave function can represent both alternatives at once.

What happens then is that such an atom is interacting with a Geiger counter. According to quantum mechanics, you come to a state of affairs in which the Geiger counter is triggered and not triggered. It is a superposition of two states of affairs: one in which the Geiger counter is triggered and one in which it is not. And if the Geiger counter is linked up to this apparatus affecting the cat, then you wind up with a state of affairs in which the cat has been killed and not killed. That is, it is a superposition of two states: one representing a live cat and one representing a dead cat.

This seemed to be a somewhat strange state of affairs. You can ask: Well, what happens if you look at the cat? According to the mathematics of the quantum theory, what happens there is that you wind up with an observer who is in many states. He becomes the superposition of an observer seeing a live cat and an observer seeing a dead cat.

This is a little diagram, which sort of represents the state of the observer. He is not looking at a machine that kills cats in this case. It is a Wilson cloud chamber that he is supposedly looking at, but a similar thing can happen there. The point is that as far as the theory goes, the observer becomes a superposition of many different observers with different experiences. So this is what is actually predicted by the mathematics of the quantum theory.

One can ask: Well, what does that mean? It seems to be a somewhat strange state of affairs.

Over the years people have had many different approaches to dealing with this. I will just indicate a few of them. One of these is the Copenhagen interpretation of Niels Bohr. According to this, you divide reality into classical and quantum realms. You restrict quantum mechanics to microscopic entities like atoms, and you use classical physics for macroscopic entities. And at a certain point you just switch over, from one theory to the next. You do that at a point when you make a measurement with some macroscopic instrument.

So this works operationally. But it seems to have this dichotomy between two completely different ways of describing nature. It does not seem to provide a unified picture of what is there.

Another one is the theory that consciousness collapses the wave function.

This is an idea developed first by John von Neumann, and then Eugene Wigner also introduced this idea. So according to this idea, as soon as an observer becomes conscious of a certain state of affairs – that is, let us say, you have a human observer looking at the cat – as soon as the human observer looks and sees that the cat is living, then you take all these different alternatives, like the alternative of the live cat and the alternative of the dead cat, and erase all of them but one. And you do that at random according to the probabilities generated by quantum mechanics.

So that is one concept. This leads to the idea that consciousness is affecting the states of matter in nature.

Another idea is that there exist many coexisting parallel universes. According to this scheme, you accept all the different alternatives as being real, and you just say that in effect the universe is constantly splitting into many universes, in which all the different alternative possibilities actually take place.

So in one universe there is a live cat and an observer seeing a live cat, and in another universe there is a dead cat and an observer seeing a dead cat, and so forth. Of course this leads to some very amazing considerations, because it turns out that this process of splitting universes continuously goes on according to quantum theory.

Actually this is a very controversial area, and I can hardly even begin to discuss all the different points. But the basic idea here is that in quantum mechanics, the basic mathematical description of physical reality tends to spread out in a higher dimensional space and represent many different mutually contradictory states of affairs. It can also represent many different combinations of events.

Here is a simple example: Suppose you let decay of radioactive atoms control the typing out of letters by a computer, so that they are typed at random according to the decay of those atoms. And let us say you let it type out 100 letters, and there are 26 letters in your alphabet. And let us say you choose it so that the probability for any one letter should be the same.

Well in that case, for the 100 letters, there are 26^{100} possibilities. So the wave function split, as it were, into 26^{100} different branches, each one of which represents a different alternative. This is where Elsasser's idea came in. He was saying that you can have a law in nature which makes a selection of one of these 26^{100} alternatives. That law is independent of the laws of quantum mechanics, because you cannot in any way measure a probability as small as $1/26^{100}$. And thus new principles independent of quantum mechanics, but not contradictory to it, can be introduced. This was his idea.

Now as far as it goes, that is interesting. What I wanted to discuss was some further developments that one can arrive at by examining this idea a little bit further. The best way to approach these ideas is to consider a simple example from classical physics. This is, say, a "potential well" that is in two dimensions. These contours represent lines of equal potential, and the lowest point is in the center.

So, you can imagine here a sort of triangular basin. And you can imagine taking a ball bearing and having it roll within this triangular basin, which acts as a force field. Or you can imagine a magnetic field, which has just these lines of equal potential.

It turns out that a simple classical system like this generates something that is called deterministic chaos, which is something that has been studied quite a bit recently. The idea there is that the path of the ball cannot be predicted for very far into the future, even in classical physics. The reason for this is that even very tiny changes in the direction of the ball will result in very large changes in its motion in a very short time.

For example, these diagrams represent computer simulations showing how the trajectory of the ball goes through a certain surface. Each point represents a measure of what the ball is doing as it passes a certain surface – actually this is done in what is called "phase space." The interesting point is that these points just move at random across this diagram. (MNS Chapter 6, Fig. 5)

Comment: You are talking about "strange attractors," right?

Answer: Yes, strange attractors are there.

In general, it involves this whole subject of deterministic chaos. One observes that the trajectory becomes as unpredictable as a completely random process.

There are many interesting examples of this. One example that is quite interesting is turbulence. This is just a computer simulation of turbulent flow of a fluid. This also has this property of deterministic chaos, which means that even though it is a classical system, you cannot predict what is going to happen for very far into the future.

It turns out that this inability to predict has implications concerning the quantum theory, because in the quantum theory there is uncertainty in the physical state of affairs of about the order of magnitude of a number called Planck's constant (h). This is about 10^{-27} erg-seconds, in those particular physical units.

So very tiny changes in position or momentum of the order of h can be amplified within a fairly short time to produce changes that are so large that you can easily see them. And the result then is that this quantum mechanical spreading out of the system into many mutually contradictory alternatives will even occur for regular classical systems, like this fluid, for example, fluid flow. This indicates, for example, that you can never really hope to predict the weather – if the weather is an example of fluid flow like this – because quantum mechanical uncertainties can actually have a large effect on the weather over short periods of time.

It is also the case that in classical examples you can show how it is possible to add laws to the existing classical physical laws in such a way that you do not violate those laws, and at the same time, you have introduced new principles which are not predicted by those laws. This is somewhat of a strange idea, if you consider the idea that classical physics is supposed to be deterministic.

This is a simple example, which shows how that is possible: If you have a series of steel pins, as in an old fashioned pinball machine, and you have one ball bouncing against them – and we assume no friction and ideal elastic collisions and all that sort of thing – then it is possible for the classical trajectory to follow any path whatsoever you might like.

For example, you can say: Let us spell out Shakespeare's plays in some particular handwriting. Well, it is possible for the classical trajectory to do that. There is a fairly simple mathematical proof of that. I will not go into it, but this leads to an interesting way of looking at the laws of physics.

Perhaps the easiest way to think of this is to go back to this example of the simple potential that I was talking about before. This is a particular trajectory, or path, followed within this potential – just one example of such a thing. You can imagine the trajectory, or path, to be an object in spacetime. And you can regard the laws of physics as describing the local connections between the different parts of this object. In the case of a simple trajectory like this, it is just the points as you go along the trajectory viewed as being in space and time.

The implication of these observations concerning deterministic chaos is that if you look at the trajectory locally, in any small part of it, you will see that it follows the classical laws of physics very closely. But if you look at it over a longer period of time, you can see that what it does is completely unpredictable.

This suggests that one can look at the laws of physics as applying to the local situation, but as not specifying what happens globally, or over a longer span of time. One can then regard the entire trajectory, say of a physical system, as a kind of continuum which can be, so to speak, flexed into any different particular configuration that you may like.

I can make an analogy here of taking a piece of flexible metal like a spring. You can bend the spring into various shapes in such a way that locally it almost exactly follows the path determined by the physical properties of the tendency to resist bending and so forth – the elasticity of the spring. But over a long distance, you can curve it so as to produce, say, some modern abstract sculpture according to your particular desire.

Of course, one would say in the case of the spring that no matter how you bend it, it will really bend quite a bit as you go along the length of it. Thus, it will not really be following its free, desired path, which would be a straight line.

But the nature of the laws of physics is such that you can view the entire spacetime history of the physical system as being a kind of elastic spring, as it were, that can be bent into many different desired shapes, and still locally it almost exactly follows the laws of physics. This leads to an approach to quantum mechanics, and also classical physics, which is another alternative to these different approaches that I outlined.

Essentially, in quantum mechanics, the problem or the situation that arises is that the system is not determined precisely. Certain aspects of things are determined, but there is a lot of freedom in the system. So, one way of looking at it is to regard the spacetime history of the physical system as a sort of flexible entity in which the laws of physics, or of quantum mechanics, determine what happens from moment to moment fairly closely, but with some freedom. And the freedom is such that the overall pattern can be chosen more or less at will.

Question: Would a naïve analogy be something like, let us say, as in differential equations, only the first derivation needs to be determined, and the rest is free?

Response: Well, let us see.

In terms of classical physics, the simplest way to understand it ... I do not want to get into technical details, but there is the idea that instead of using differential equations to describe the trajectory of the system, using the idea of finding a stationary point of the action. This action for a trajectory is a global description of the trajectory; it involves an integral of local events over the total length of the trajectory.

And there is a classical principle, I guess originally developed by de Maupertuis and also similar to Fermat's principle in optics, according to which the path of a light ray through a series of lenses can be determined by considering the refraction of each lens. Or you can consider the total time it takes for the light beam to go through and take that path which minimizes the time.

So let us just take that light beam example. It might turn out that as you vary the path which will minimize the time passage, you find that you are in a very steep valley. That is, as you vary this path in this multi-dimensional space of all the different parameters that you can vary, varying it slightly takes you way off from the minimum. If that is true, then the path is very well defined, it is very exactly determined.

On the other hand, suppose as you vary the path you are in a very broad valley in this multi-dimensional space – it is almost flat. In that case, it could be that you could have many different paths which look almost equally good.

The main implication of these findings in deterministic chaos and so on is that in many physical systems it is like that. The valley, so to speak, in this multi-dimensional space determining what is the proper path, is so flat that all kinds of different paths are possible. And quantum mechanics makes this situation even more the case. It actually, you can say, makes the valley even flatter.

We can then think of this idea of the total spacetime history of the physical system as being very freely adjustable, and yet, the laws of physics are locally obeyed. So if you do experiments, you will always see that the laws of physics work out, as you would expect.

To go back to this idea of Elsasser, this is a specific way of introducing his concept of "creative selection." The idea here is that you can take this spacetime continuum, and so to speak, working on it from a vantage point that is transcendental to time – that is, that you are taking the whole history at once, and molding it – you can get it to produce all kinds of different forms, for example, structures of organisms and so on and so forth. At the same time, this whole bending and flexing procedure produces a pattern that is completely consistent with the laws of physics.

This is a possible model which can be applied to Elsasser's idea. I would like now to go back to this concept of consciousness that I mentioned before.

I will mention, in particular, an interesting experiment in parapsychology, which has been reported in recent years. The original work was done by a man named Helmut Schmidt, a physicist. He had a radioactive source which controlled a random number generator that utilized a circle of lights, and only one light would be on. The light that was on could either shift to the right or the left, with a fifty-fifty chance, depending on the radioactive decay.

When you looked at it, you would see a light that would sort of be randomly moving around in a circle, back and forth. Now according to quantum physics, with a fifty-fifty chance, this light should just do a random walk. All the statistical properties of a random walk should be there.

But he then would ask a person to look at this light and try to will it to go around clockwise. And he found that at least with some subjects, over a period of, say an hour or so, they could consistently make it tend to go around clockwise. It is not that they could just make it march directly in a clockwise direction; it would be going back and forth. But on the average, it would keep going around clockwise.

Also, there were some people who wanted it to go clockwise, but on the average, it would go around counterclockwise. Things seemed to be happening in a perverse fashion for them.

So, this experiment is interesting. This is an example of a parapsychological experiment involving psychokinesis, and of course that is a whole subject to go into.

But the interesting thing here is that you can ask: What is it that is being influenced that makes the light tend to go around clockwise? If indeed all this is correct, and it is really doing so, it would seem that the person is not changing what is going on directly within the arrangement of lights, nor is he changing what is going on in the circuitry that directly controls the lights. It would seem that the change must occur at the level of the decay of the atoms. In fact, Schmidt checked this by monitoring some other stages of the equipment and so forth.

So then you can ask: Well, how could the subject who was willing the light to go clockwise possibly know what the atoms were doing, or how the decay of an atom in a particular way would influence the random number generator so as to make the light go in the desired direction? Because the subject who was willing the light to move around, of course, was not aware of how the whole apparatus was built.

In fact, it is possible to make the apparatus quite complex in such a way that even if you knew how it was built, and you were able to control the decay of the atoms, still it would be hard to do so, so as to get the desired result.

What seems to be involved here is that we have events occurring in such a way that they anticipate the will of an individual. And this seems to involve a violation of causality. That is, because the individual is willing something to happen at a given time, at an earlier time, something that he does not even know about is happening so as to enable that desired thing to happen. This seems to be quite a paradoxical state of affairs.

It turns out, though, that this particular way of looking at physics is quite compatible with this kind of event. Let us just imagine, for the sake of argument once again, the space-time history of events as a kind of flexible continuum. And let us imagine that the will of an individual could have some influence as to what state that continuum was in at a particular time, so that the continuum could be drawn to go through that state.

Well, if locally the laws of physics must be satisfied, and the continuum can be somehow made by will to go through that particular state of affairs, representing, say, clockwise motion of the light, then it follows that everything has to line up in the time preceding that state of affairs in such a way that at each stage the laws of physics are satisfied. And then it follows necessarily that the atoms are forced to decay in such a way as to ultimately produce that result.

In other words, if we look at the laws of physics in this way, we find that this kind of anticipation of the future, or apparently teleological behavior of physical systems, is quite compatible with the theoretical structure.

I wanted to conclude by observing that this particular model, which one can tentatively think about, is just a model being proposed for thoughtful consideration. And this particular model is also consistent with some of the ideas in the Vedic literature concerning the nature of the relationship between consciousness and matter.

Briefly, in the Vedic literature there is the concept that essentially there are two kinds of conscious entities: the individualized conscious beings, such as ourselves – the term *jīva-ātmā* in Sanskrit is used to refer to these entities – and also there is a super-consciousness, which is simultaneously conscious of all phenomena and is directly in control of all phenomena.

It is described in Vedic literature that this super-consciousness, or *Paramātmā*, relates with the material realm in a manner that transcends time and space. Essentially, on the level of the super-consciousness, or the level of *Paramātmā*, there is no difference between past, present, and future. And also, all places, all spatial locations, are simultaneously observed. The idea is that this one super-consciousness perceives the entire space-time picture at once and can control that through the exertion of its will.

The Vedic literature also describes the nature of the relationship between the individual conscious entity and this super-consciousness as being one in which a conscious entity associated with a particular body is not directly able to govern the actions of that body. This for example is described in the *Bhagavad-gītā*, that the individual soul is not the performer of actions within the material world.

But, the idea is that the link between the individual conscious entity and the material energy goes through the super-consciousness. That is, the super-soul is aware of the desires of the individual entity, and the supersoul controls the material energy in correspondence with those desires. But not, of course, exactly in correspondence with them, because different conscious entities may have incompatible desires. A rich man may desire that his wealth will not be stolen, while a thief may desire to steal it. So one of them will not get his way.

The idea then is, in the Vedic literature, that the supersoul, acting in a global fashion, determines the sequence of events within nature in such a way as to optimally satisfy various desires and so forth of individual conscious entities.

So this is also a very complex topic. But, one can see how that would fit into this picture that I am describing. We can look in particular at this parapsychological experiment. In this case, if the subject of the experiment, who is looking at this arrangement of lights, is desiring the lights to go around in a certain way, then, according to this model, the supersoul can adjust the space-time picture in such a way that it happens. And that also involves events which anticipate the movement of the light.

Actually, Schmidt performed an even more disconcerting experiment. He had the random number generator produce a tape about a week before the person was to look at the lights. Then instead of using the radioactive source to run the lights, he used the tape to run them.

And he again got positive results. The person willed the light to go around clockwise. But yet the specific sequence of ones and zeros, which determined what the light would do, had been recorded a week before.

Question: The question is, who is plugging into what? Is the man plugging into the pattern that has already been set by the computer? Or does the computer just include the man plugging in?

Response: You could ask: Well, could it be that because the pattern came out the way it did the week before, therefore the person decides to will to make it go a certain way?

Of course, one problem with that is that you would not expect the pattern to go around consistently in a certain way. You would expect it to do a genuine random walk, in which on the average it wouldn't go around in any particular way.

And then even if you suppose that the person later willed something because of what happened earlier, the question is: What is the connection that caused you to do that? So, these experiments involve many problems of interpretation.

But the observation that I thought was interesting was that this general model of the relation between consciousness and matter is consistent with that. And of course, another feature of this model is that this model enables you to preserve the laws of physics, but it still allows a superconscious being to control in considerable detail the course of events.

Well, I have spoken for about an hour. I will stop here. Are there any questions or comments?

Question: You did not really mention the similar observations and research by Richard Cartwright [*name unclear*] based on the holographic paradigm. He is also trying to establish models, where they are observing that at a certain point matter is no longer matter. The pattern is there, but the matter is no longer there. [*name unclear*] of course, that the dance is without the dancer, where there is the dance but there is no longer a dancer.

Therefore, matter can at that level be directly influenced by the [*unclear*] observer who is observing matter, which seems to be the point where matter, or the universe, is being created by mind. They are very much in agreement on that level of physics, that there is something there that needs to be taken into account, at global levels as well, if you want to escape the paradoxes that appear at the time-space continuum. But when you go beyond that, the paradoxes are no longer there.

Are you familiar with that?

Response: I am somewhat. I am familiar with the writing of David Bohm in particular.

Comment: Right, yes.

Response: Bohm has introduced some ideas along these lines.

Actually, one starting off point for me in considering this was some of Bohm's examples of his "implicate order." In a way you might say that what I am presenting here is a concrete version of what Bohm is saying, with I think, a couple of additional features.

One interesting question is: Let's say you have the implicate order in which things are very much mixed in with one another, in his idea, giving rise to an "expletive order," which corresponds to the realm of reality as we observe it. Then there is the question of how the implicate order gets determined to be in just the right way so as to produce this kind of expletive order. This is sort of like the problem of initial conditions.

Consider for example, this bouncing ball that I had before. It is indeed a fact that you can choose the initial angle of the ball so the ball spells out Shakespeare's plays. That means that in the number representing that angle, all of Shakespeare's plays are present. That is sort of like this implicate order situation.

It is also very much like a hologram. In a hologram, in one little piece, an entire picture is present. Similarly in this one little number, you have all of Shakespeare's plays stored up.

But you can ask: Well how is it that it would come to be that this angle is just right so as to give you Shakespeare's plays? You need a really good aim for that. But the approach here is that from a global perspective, you can say:

Okay, let us determine this to be Shakespeare's play. Then it follows that the angle would come out right.

So possibly this is a contribution to those ideas, in that it explains how some of these initial conditions, which I think are entailed in Bohm's system, can turn out to be so precisely specified so as to give rise to so many different phenomena.

Question: Does it mean that there is a difference between the cause of one act of creation, or is it a continuous process of creation that keeps adjusting?

Response: Well, the scheme that we have here is one that you can look at either from the point of view of progressing through time, or from a global point of view that transcends time.

According to the Vedic literature, the platform on which the supersoul actually operates is global and transcends time. But then you can project that down into the level in which time is passing from moment to moment.

According to that scheme, from looking at it from the point of view of time passing from moment to moment, you should actually see events which act in such a way that something can happen later. And it might be very hard to see how they got aimed just right. How the initial conditions were selected properly. But from a global perspective, then it makes sense how that could happen.

At least that is one insight that I think perhaps can be gathered from this model.

Question: Would you have put [*unclear*] limitation of speed of light propagation changes, etc.?

Response: No, it turns out that this particular model, this way of looking at it, is consistent with the theory of relativity, which is interesting. Because what it says is that propagation within space and time can only go at the speed of light.

But you can have trans-causal signal propagation if you like. That is, according to this scheme, it is possible for the space-time continuum to be bent or molded in such a way that you actually have correlations set up between two points such that a beam of light could not travel from one to the other. That is, they are outside of one another's light cones as far as relativity goes. But that does not violate the theory of relativity, because transmission within space and time is still limited to the speed of light.

It is also interesting that this point of view avoids some of the causality paradoxes that you seem to get into when dealing with things like tachyons and other hypothetical particles, which by traveling faster than the speed of light can go backwards in time, and so forth.

Comment: I will just ask you – a few years ago, I think 1980, there was a research paper published by, I guess his name was [Robert] Jahn. He was a dean of aeronautic engineering at Princeton University. It was published in [*title unclear*] Proceedings, with his experiments in parapsychology.

Do you recall? I never had any reaction to it, but it was quite interesting that somebody of that high position who earned many international awards, etc., would have association with the subject. But I never had any reaction.

Response: Well, I went and visited his department there at Princeton. I did not have the opportunity to speak with him personally. I spoke with another professor who was working there. They have a laboratory in the basement called "Laboratory for the Study of Engineering Anomalies." That is how they term their research into parapsychology.

They are doing extensive research into things like psychokinesis, and they say they have obtained very good results. They are using computers and making statistical studies and so on and so forth. They have apparently corroborated the kind of experiments performed by Helmut Schmidt, for example. Other investigators have also corroborated these things. Of course, many scientists will shy away from this kind of subject matter. Although, I think an increasing number are beginning to take an interest in it.

[*break*]

Comment: Well, in fact, I did not understand where the extra ability [*unclear*] comes from. I have the impression that you mixed up the laws of quantum mechanics, say the equation of the [*unclear*], and the laws of classical mechanics, the Newtonian equations.

If you consider the Newtonian equations, then the past of a particle is completely fixed by its initial position. It is completely fixed for all times. I have the impression that you added a small deviation, say *epsilon* > 0 [*not entirely clear*], and a small deviation has a quantum [*unclear*] ...

Response: We can look at it in this way.

Yes. you say, in a classical system, if you specify the initial conditions, the entire trajectory is completely fixed from that point onward. But the point is then that exceedingly small changes in the initial condition can affect the trajectory in a very drastic way. So much so that it can become completely different just based on a very minute change. In fact, the change postulated in the initial conditions can be so small that it is on the level of quantum mechanical fluctuations.

Now, this is a sort of loose way of going from classical to quantum mechanics. To do the thing properly one has to analyze the Schrödinger equation and so forth. I did not try to go into any of that.

Actually, the most appropriate way to approach the quantum mechanical aspect of this problem is to use the Feynman path integral approach. The reason for this is that the best way to analyze classical physics in this context is to use the action minimization principle. And the Feynman path integral is the quantum mechanical analog of that. So, if you then go to the Feynman path integral, that is the way to approach things.

But, in this discussion, in a sort of loose way, I went from classical to quantum mechanics. Simply by the idea that if you have a fluctuation of quantum mechanical magnitude – that is around h or thereabouts – being amplified in a fairly short time to produce a really big effect, then one would expect that if you actually solve the Schrödinger equation for the system, you would find that the wave function, in a very short time, would spread out to encompass many alternatives.

For example, in this bouncing ball case, by slightly changing the initial angle the path will be radically different. So, one path will spell out Shakespeare's plays, let us say. And of course, most paths will just be chaotic and go all over the place.

This would suggest that if you really solve the Schrödinger equation for that case, then the wave function describing the path would very quickly spread out so much as to represent all these different paths. In fact mathematically you can show that.

Comment: Well, of course in a sense, also, quantum mechanics is initially … when you hit *psi*, it is also fixed. *Psi* is fixed.

Response: Right. *Psi* is fixed. But *psi* encompasses many alternatives, such as live and dead cats.

Comment: Yes, that is true, But then, of course, comes the theory of measurement into a [*unclear*] …

Response: Right. And then there are different approaches.

So, one approach that has been taken is to say: Very well, *psi* spreads out, so we will not stick with one *psi*. At a certain point we will collapse the wave function. We eliminate all the alternatives but one. And then that is the new *psi*. And then that spreads. And then we have to do it again. And again, collapse the wave function.

Now this leads to problems in interpretation. For example, some physicists such as Eugene Wigner will say that conscious perception of the observer causes the collapse of the wave function. Other physicists do not like that at all.

There are many different proposals that have been made. One of these even is that: Well, let us not collapse the wave function, and let *psi* as a whole represent reality. Then you have the multiple universes scheme in which there is one universe in which this conversation is going on, and another one in which we are doing different things, and another one in which we do not even exist, and so on.

All these are attempts to solve the problem of measurement in quantum mechanics. If you like, you could say the thing that I am presenting here is another one, another approach which has, I think, some interesting features.

Comment: You are mentioning that at one precise angle you get all the words of Shakespeare, and then when you go off that angle, you get chaos. I do not know whether I object to that, in the sense that, it would seem to me if one angle produces all the words of Shakespeare, at that moment you determine it to be Shakespeare because you recognize that it is Shakespeare.

That would seem to me that every angle would produce its own path, very precisely and very much repeatable. So, it means that every pattern would be as complex as the words of Shakespeare; simply that you do not have the vantage point to actually recognize in fact that it is better [*unclear*].

Response: That brings up a very interesting question, because if it does produce the works of Shakespeare, then we will say that it has produced a meaningful statement.

Comment: Right.

Response: But let us say it produces something else. Could there be some kind of intelligent being to whom that would be a meaningful statement? Could it be that for every possible complicated zig-zagging path, there exists some intelligent being for whom that is a meaningful statement?

Comment: Well, it belongs to someone, some author who has produced the work [*unclear*] of Shakespeare. It would be recognizable by someone else.

Response: Right.

Is there a language, such that if you knew that language, then you could read it, and it would be the works of that author ... [*break in recording*] ... compared to our own worlds, somewhat along the same lines, but quite alien – perhaps in which you have octopuses, as in the science fiction stories, talking with each other, and so on. Or worlds that we do not even know how to think about, because they would be so alien.

So all these possibilities are there. Then what selects this world?

Comment: Same vantage point, rather than angle.

Response: Well, vantage point is included within the world, of course. Because it is one thing to just look at the balls, and another thing to have an outside observer who is going to read this as Shakespeare.

But now if we look at a larger picture, which includes the observer, and we start out with some initial conditions that result in observers observing certain things, then the vantage point is also within the observed system. So how many different arrangements of the initial conditions would produce systems in which you have observers, who will say that I observe such and such?

Comment: Yes, but I thought that in epistemology you stop at a certain level. You stop after that level, and you are watching the watcher, watching the watch. You watch the watcher that is watching an object. After that, it makes no more sense to go on exponentially.

So, it seems to me that you are describing basically, you have an individual. The vantage point is watching the pattern that is being created by a certain angle, and after that it is *Paramātmā* [super-consciousness] that is doing the watching of all the influencing [*unclear*].

Response: Well yes, of course that is the basic scheme I am proposing. That is the basic idea.

Comment: But the point is, as individual consciousness, can you get to the level of *Paramātmā*? Can you get to the state where you are watching the watcher, and the watch?

Response: Well, can we get to it? *Paramātmā* is at it already [within Vedic philosophy].

So that, of course, is a fundamental question.

According to my understanding of the Vedic literature, the individual consciousness remains individualized. The consciousness of *Paramātmā* is always the universal consciousness, and the individual consciousness, inasmuch as it is an individual consciousness, is individual. If you say by some means it can become on the level of *Paramātmā*, well, then you have got *Paramātmā* and the individual is not there anymore. In other words, in one sense logically, the individual has been annihilated, and now you have just got *Paramātmā*, which you had all the time anyway.

So the question then comes – does the individual exist eternally, or can the individual cease to exist at a certain point? And that is a topic for discussion in the Vedic philosophy.

For example, my understanding of the *Bhagavad-gītā* is that it would postulate individualized consciousness as being eternally coexistent with the consciousness of the *Paramātmā*. So that the ultimate stage of realization of the individual is realization of their relationship with the *Paramātmā*. And according to this concept, ultimate oneness means ultimate agreement, or harmony, with the will of the Supreme – not becoming identical with the Supreme, since that would actually, in a sense, be equivalent to annihilation. Because it would mean then the individual is no longer in the picture, and the Supreme is the only thing that is there.

So, that of course, is a whole subject to discuss.

Comment: What you are essentially saying is that the theory of creative selection is a way to look at quantum mechanics such that there can be no contradiction with the Vedic philosophy. Is that correct?

Response: Yes, you could say that that is more or less what I am synthesizing here, putting together.

In other words, and I do not propose this particular way of looking at quantum mechanics as being the last word or anything like that, but just as an interesting exploratory attempt. The idea here is that one can take the laws of quantum mechanics and interpret them in such a way as to be consistent with the idea that there is a supreme intelligence controlling all phenomena. This, of course, is relevant to some very long-standing questions in theology, for example.

When Newtonian physics first came on this theme, the old idea of divine providence, according to which God is in control of the motion of every blade of grass and so on, seemed to become now inconsistent with physics, because it seemed that the laws of classical physics are completely deterministic. So then if God had any role in actually determining what happens, that role had to be exercised at the beginning. God could set everything in motion, and then according to the laws of physics, it could all work out.

So, according to this way of looking at quantum theory, it is possible to have the laws of physics as observed by physicists and so forth, but it is also possible to have God in control of everything. So that is more or less the main idea of this.

CHAPTER 10

A Trans-temporal Approach to Free Will and the Laws of Physics

Notes:

Chapter 10 offers two presentations. The first, (10a), is a transcript of a talk Richard Thompson gave at the conference "Consciousness within Science" held in San Francisco on February 17–18, 1990. Section (10b), a facsimile reprint of an earlier manuscript, offers a formal analysis of many of the arguments presented in (10a).

The 1990 conference at which (10a) was presented had been hosted by the San Francisco branch of the Bhaktivedanta Institute and was attended by a number of prominent scientists and scholars including the 1963 Nobel laureate John Eccles, as well as Karl H. Pirbram, Henry Stapp, and John Searle, among others.

While Thompson's 1990 presentation appears to draw from research work analyzed in Section (10b), Thompson offered a less technical presentation on these themes in a talk given at the January 1986 "First World Congress for the Synthesis of Science and Religion," also hosted by the Bhaktivedanta Institute, in Bombay (Mumbai), India. This event had been similarly attended by eminent scholars such as George Wald (1967 Nobel laureate), the Dalai Lama, Harvey Cox, and Huston Smith, among others. A published version of the talk with the title "God and the Laws of Physics" was included in the conference proceedings, *Synthesis of Science and Religion: Critical Essays and Dialogues* (1988), on pages 213–237. Thompson reprinted the essay with the same title as Chapter 1 in his compilation volume, *God & Science: Divine Causation and the Laws of Nature* (2004).

Following the 1990 conference presentation (10a), there is a set of images related to the event.

CHAPTER 10A

Free Will and the Laws of Physics: A Trans-Temporal Approach to Mind-Brain Interaction

Presented at:
"Consciousness within Science"
Bhaktivedanta Institute
San Francisco, CA / February 17–18, 1990

I thought that I would begin this talk by offering a solution to the problem of free will.

In fact, I will begin by giving what I think might be John Searle's solution to the problem of free will. He will be able to correct me later to see if I have got it right. Actually, when I heard his presentation on consciousness being an aspect of matter, I was a little surprised because I thought actually this was a very common idea.

For example, there is a friend of mine – a biologist – who expressed this idea in terms of what he calls "hylozoism." According to his concept, consciousness is matter as seen from the inside. Atoms, molecules, and all that, that is matter as seen from the outside.

That is actually a fairly common idea. So, in the context of this idea about matter and consciousness, what can we say about free will?

Well, according to this idea I would have to say that I am completely a machine, operating according to the laws of nature. I am this body; this body is made up of little parts and particles. These parts are interacting according to certain rules given by the laws of physics. And that is the total story.

If we accept that view, then what can we say about free will? Well, the answer would be: There is no problem of free will there. I can do what I will, including lift my arm and so forth, because after all I am a machine. What the machine does is what I do. There is no question of any absence of free will. We are simply identifying states of consciousness with states of matter, and the problem disappears.

So that is one solution. Or at least you could call it a solution. However, there are some who may not be fully satisfied with this particular solution to the problem of free will, or indeed, with this solution of the question of the relationship between consciousness and matter.

Actually, another good point that John Searle made is that it is a question of the facts. What really is the relationship between consciousness and matter? We may be satisfied with a given theoretical viewpoint, but the facts are really the important thing. And what do we do if we are not sure about the facts, or if at the present time we cannot really come to a consensus in which we can all agree about what the facts might be? Well, in such a situation, we should at least be open to looking at alternatives to consider different possibilities.

There is, of course, another possibility to this idea of identifying consciousness with matter, and this is basically the idea of some kind of dualism. The idea in dualism is that there is mental, or conscious reality, consisting of feelings, states of intentionality, volition, emotions, perceptions of color, of pain, of desire, and so on and so forth. All of that is in the domain of consciousness or mind. And then there is matter as we know it, which we describe in terms of atoms, molecules, quantum fields, and so on and so forth. So, this idea of duality is there.

Now in the context of this idea, what can we say about free will, or the concept of free will? Well, given a dualistic perspective, which, for all we know at the present time may be the fact of the matter, the question becomes: How can you have an interaction between consciousness, or mind, and matter?

The idea is there that when I lift my arm it begins with volition, which is an aspect of consciousness or mind. Within my mind I desire to lift my arm, and then the physical mechanism goes into operation. And of course we are quite sure, at the present stage of scientific development, that the brain is very much involved with this. My arm moves because nerve impulses went to the muscles. The nerve impulses originated within some cells within the brain. It becomes a question of interaction between the brain and this conscious mind.

Now that immediately leads to some serious questions: How in the world can the mind – conscious mind, if we postulate that there might be such a thing – interact with the brain? We run into problems involving the laws of physics. The basic problem we come up with here is the following: The laws of physics seem to give us seamless continuity in the patterns of cause and effect which determine all different phenomena occurring within nature. There is no convenient gap in the chains of cause and effect.

In classical physics for example, we can express the laws of physics in terms of differential equations in which you advance the physical states of the system going by infinitesimal increments of time, using an equation to determine how things change from increment to increment. This is deterministic, and there seems to be no room for introducing any kind of intervention of will or consciousness, volition, within the physical system.

And it does not get any easier if we go to quantum mechanics. Now in quantum mechanics we introduce the idea of chance. As Henry Stapp was telling us, in quantum mechanics you have in the typical formulation a wave function which represents – at least Heisenberg took it to represent – potentiality.

This wave function represents different probability amplitudes for different configurations of matter. Essentially, for all the different waves, matter might be combined, and you have a complex number, which is a quantum mechanical amplitude. As time passes these amplitudes change according to the Schrödinger equation, which is a deterministic, mathematical equation. The amplitudes are changing continuously in time, but they come to represent different macroscopic states of affairs. This has become famous as the "Schrödinger cat paradox."

The set of amplitudes can diverge, so that you can have non-zero amplitudes representing the configurations corresponding to a live cat and to a dead cat. Your quantum mechanical description is simultaneously telling you that you have got a live cat and a dead cat in one and the same place. It is one and the same cat.

So this seems very difficult. This is one of the mysteries of quantum mechanics that has been debated since the very early days of the quantum theory.

So in quantum theory, what do you do about that? There are two approaches to this, as Henry Stapp was pointing out – epistemological and ontological – and I am going to discuss these matters more a little bit later on. But basically, you have to make some choice. And you make the choice by taking the square of the absolute value of your amplitudes to the different configurations you have, normalizing them appropriately, and interpreting those as probabilities. Then by pure random chance you pick a solution.

That is what you have in the case of quantum mechanics. You have a stochastic process.

So there also, it is a bit of a problem to introduce any concept of free will into the system. The reason is, you might say: Well, we have freedom here, because there are different options that the system can follow. And the system collapses to follow this option, or it collapses to follow that option. But, it does so at random.

In the collapse of the wave function, which occurs stochastically, or in a random way, it is also hard to see how you can bring in free will. Consider the idea of tossing a coin simply by chance. Let us say it is a fifty-fifty chance heads or tails, so you have independent, identically distributed random variables, describing how the coins are going to go.

Well, what free will is there, *there*? The outcome occurs according to deterministic rules of statistics, which give you the rule that in the long run you are going to have fifty percent heads and fifty percent tails. You need the law of large numbers there. The sequence head-tails will occur about twenty-five percent of the time. The sequence heads-heads-tails-heads would be $1/2^4$ percent of the time, and so forth.

So that is a completely statistical process, and there is no question of free will unless you want to simply say: Well, that is free will! It really does not correspond to the idea that the mind, a conscious entity …

[*short break in recording*]

… value of size squared, the amplitude squared. Then what becomes of quantum mechanics? It is not quantum mechanics anymore. What becomes of the very meaning of the probability amplitudes? They become meaningless. You have got a totally different theory.

It is a real problem to try to introduce any idea that freedom of the will on the part of the mind, the conscious entity, can influence the effects going on in nature within the framework of the quantum theory as it stands. That is also true for the classical theory; it is true for the quantum theory. Those are all the theories we have.

So how are you then to introduce this idea of free will into the modern scientific picture of the world given to us by physics? I would propose that in order to do it – if you want to do it at all – you would have to introduce some new physics. There is really no way around it.

Now the question is: What sort of new physics should you introduce? There are various levels of change you could consider, or contemplate making, in the laws of physics. Very prominent physicists have contemplated such things. For example, Eugene Wigner proposed replacing the collapse of the wave function with some nonlinear function, which represents the action of consciousness. I do not know if he ever carried through that program in mathematical detail, but that would be quite a radical change in the whole theoretical structure.

Then of course, there is also the question of the magnitude of the changes that you would postulate in the action of matter. Just consider the history of the action of matter as time unfolds.

We can measure to what extent that follows the laws of physics. The way you could do it – this is simplistic – is to allow for some really big deviations to occur, something really radical. The laws of physics say it goes like *this*, and the actual fact of nature is, it goes like *this*. And that would be a really radical change.

There is also the question of conservation of energy. Perhaps the kinds of changes you would need to make in the laws of physics, to allow consciousness to have a control over the evolution of the system, would involve a violation of conservation of energy by some measurable amount. That is another idea.

This is one kind of change in the laws of physics that you might contemplate; that is one alternative. Perhaps it is even true that these things happen; this could be considered. But what I am going to do in this talk is consider another alternative to how you might modify the laws of physics. According to this alternative, you only make exceedingly small changes.

So, can we do this in such a way as to allow consciousness, in a dualistic concept of mind versus matter, to control a physical system? In order to explain that, I want to introduce a few ideas, and I made a few pictures.

Here is a general picture for you to visualize in the course of this talk. This is a total spacetime picture of what you have in classical physics. Basically, what we have is time on one axis here and configuration space on the vertical axis. Configuration space means all the coordinates defining the situation in the physical systems. You could have many Avogadro's numbers worth of coordinates in a realistic system.

Then there is the path, or history, of the system. This path indicates what happens in the course of time as events unfold within the physical system. This is a classical picture, and I will show you a quantum mechanical picture a little bit later on.

The first thing I want to do is talk about a new development that has occurred within classical physics. This is something that has been discovered recently by numerical studies. Somehow by analysis mathematicians missed it for many centuries, at least since Newton. This is the idea of what is called "deterministic chaos."

Deterministic chaos is a common name which is used, but the real term that one might want to use for this is "exponential amplification of very small effects." What is discovered is that in nonlinear systems in physics, if you make an exceedingly small change in some parameter within the physical system, you can have an exponential amplification of the difference between that, and what it was before, so that the path followed by the system changes radically within a very short period of time.

For example, you could change something in position by 10^{-30} centimeters. This is a classical system – we are not talking about the uncertainty principle here. You change it by 10^{-30} centimeters, and in a millisecond the thing could be moving off by one centimeter. This kind of thing can happen in what is called "deterministic chaos."

It turns out that deterministic chaos is an extremely common feature of classical physical systems. I have listed a few here. For example, a pinball machine, which is significant because we just saw a picture on the screen of a kind of pinball machine: atoms in a liquid, that is like a super pinball machine with atoms bouncing together. Turbulence, and many other examples, can be given.

What this does is suggest one immediate naive idea of how you can have consciousness controlling a physical system, without really violating the laws of physics, by too much.... [*laughter in the audience*]

The answer is: make these very *tiny* changes in a system in which there is deterministic chaos. Let the natural exponential amplification, determined by the nonlinear laws of the classical system, amplify the effects to a large scale, and there you are.

Actually, this idea of deterministic chaos is very interesting in terms of the laws of thermodynamics. As you may know, the second law of thermodynamics says information is always being lost in the physical system. Actually, it is not really always being lost. Macroscopic information is continuously being reduced to microscopic, unmeasurable, information. Because the counterpart of exponential amplification is exponential damping, just the opposite effect. So, the second law of thermodynamics is saying: macroscopic information is continually becoming microscopic so that you cannot measure it anymore. Thus, it is lost.

Well, the counterpart is, that if you have microscopic information that is there that happens to be of an orderly nature, then that can amplify and produce macroscopic orderly effects. All you have to do is explain how you could get the microscopic, that is atomic level, orderly information into your system.

Actually, you can do mathematical simulations with systems that have deterministic chaos. Normally, the problem with this microscopic information is that it is random. It is simply a random sequence of tiny pulses. The result is that when it amplifies to a large scale, you just get a large-scale random effect. And that is where the name deterministic chaos comes from.

So, the problem is, how could you get the orderly information? If you postulate that the conscious self can inject, on an extremely tiny scale, orderly patterns of information into your system, then you can get large-scale systematic purposive action according to the will of the conscious system. This will not violate laws of conservation of energy by any measurable degree. We are postulating extremely small changes that you could not measure.

It will, however, violate the second law of thermodynamics. Because instead of random information being amplified up, you are saying it is non-random. This violates the second law of thermodynamics. In fact, I think it would be fair to say that if you can at all suppose that the conscious self can exert control over a physical system, that must violate the second law of thermodynamics, at least to some extent. That is a price you have to pay.

This is a model showing one way in which you could have the conscious self controlling matter, without having to make some really drastic change in the system, like moving something substantial, and thus violating the conservation of energy drastically.

But there is a problem with this model. And the problem very simply is this: You would have to postulate that this consciousness has super intelligent, computational power on the level of Laplace's imaginary super calculator, who can predict the future of the universe; the reason being that it is very hard to determine, mathematically, how you would have to introduce these tiny perturbations in your system, at one time, so that at a later time you get the desired macroscopic effects.

It is a really difficult mathematical problem to do that. And you would have to postulate that somehow the conscious mind is solving that problem all the time, in order to manipulate matter. This would be a very awkward model.

So, I propose a solution to that. And in order to introduce that solution, I am going to bring in another concept from physics. This is the idea of "global nonlinear optimization." It is a general numerical technique. In this technique, you have some system in which there are a lot of constraints on a solution. These constraints may conflict with one another. And what you want to find is the best solution that satisfies all the constraints as well as possible, although you cannot satisfy them all perfectly.

I give here a very simple example, which I will [also] refer to later, of a spring with some rubber bands and nails. You can see here what would really happen in nature with this model. The rubber bands contract, the spring is flexible, and it relaxes into something that satisfies all the constraints as well as it can. The rubber bands want to contract completely and so forth, the spring wants to stay straight, but then you get a compromised solution. And there are numerical techniques by which you can work out what this would be.

What I want to introduce is this concept of global nonlinear optimization into the physical picture. To do that, let me go back for a moment here to this picture in classical physics. It turns out that already in classical physics we have something very much like this concept of global optimization, and that is called "the principle of least action."

This is a very old idea in classical physics, but basically, there is a function representing constraints on how the physical system should move. And if you optimize that function globally over the whole history of the system, between times A and B let us say, you get a solution which is the classical path, the classical physical path that that system would follow.

The proposal that I would make then, is simply the following:

We will introduce the idea of "will," or volition, into the physical system in the following way. In this concept, we have to regard volition, or will, in a dualistic fashion. Volition is not part of this path that is drawn here. The path represents the spacetime history of the physical system. All the ordinary interactions of cause and effect occurring within the physical system are represented within that path. We want to represent how volition on the part of the mind, the conscious entity, could change the path.

The proposed solution then is the following: Let us suppose that at different times in history, we have different volitions. Now these can be volitions occurring in different people, and of course, one person wills many different things, at many different times. So I would have to have millions and billions of "w"s here – I have just drawn three for simplicity. The idea then is to do the following thing: For each volition we shall suppose that within the brain of the individual who is "willing," there is a pattern corresponding to what he would like to have happen.

Now how this all works out in terms of mechanisms of the brain is a very, very complicated issue, and I am not even going to try to address that. But the idea is that the mind is different from the brain, in this picture. The mind has its idea of what it wants. And there is a mismatch between the pattern that is in the brain and what the mind wants. According to this idea, that represents a constraint on the system at that particular time.

The "will" in this model is that the pattern within the brain would approximate, as closely as possible, what I want. The idea is the action of "will" within the system is to try to minimize that. What we do in this global picture, and this is where the title of the talk comes in, is propose that the different "wills" correspond to constraints. The laws of physics, as given by the classical action principle already, correspond with a set of constraints. And what nature does is do a global optimization on that, to give you the actual path.

Now let me explain a little bit about what this implies. First of all, the key point to note is that because of the phenomenon of deterministic chaos, the physical system is actually very flexible. You might think that if you apply an extra constraint to the path of the physical system, it may have to be a very powerful constraint in order to get that physical system to shift so as to follow that constraint. However, this phenomenon of deterministic chaos has the effect that the system is very flexible. It is easy to get it to satisfy various different constraints.

Let me give you a simple outline of what could happen, let us say within the brain, when you will to move your arm. And once again, I am not at all going to try to talk about what happens with the different neurons and synapses. It is a super complex system. But one thing to say for sure about that system is that it is highly nonlinear, practically everywhere, within the system.

What happens then is this. At a given time you "will" to do a particular thing. According to this theory, the spacetime system responds to that "will" by giving you a path, so that at the time that you will it, the behavior of the system approximates as closely as possible to what you wanted.

Since the system is continuous and has to satisfy the constraints imposed by the action principle in classical physics, it is not that the system goes along in its ordinary way, and when your will comes in, suddenly, "bang," it jumps up like that. That would be a gross violation of the laws of physics.

Rather, it continuously moves up. What happens in effect is you have this amplification of very tiny effects. But the very tiny effects are not in this case produced by the conscious self, somehow figuring out what they ought to be and making them happen, so that then the system behaves in the desired way. It all happens automatically.

In fact, if you gave me the mathematical problem – how to determine what the very tiny effects would be which would give rise in a given set of nonlinear equations to a particular thing at a particular time and particular configuration – then I would try to solve it using a global nonlinear optimization method. That is the method I would use on a computer if I wanted to solve that problem. What I am proposing is that nature does that.

Of course I should address the objection here that nature would not do that by a process of following an algorithm, as we would do on a computer. Rather, I am simply proposing that the actual path followed in nature corresponds to what you would get by performing such an optimization.

What is the point of all this? What I am trying to outline here, and show, is that it is possible to consider that a conscious self, distinct from the material system, could influence that system in such a way as to make things happen within the system according to its will, and at the same time, not do much violence to the laws of physics. What I essentially am proposing here is, (1) take into account the idea of deterministic chaos, and (2) modify the principle of least action by adding these extra constraints corresponding to the volition of the system.

Now in terms of the actual details of what may happen within the brain, I would point out that this is fully compatible with many, many different models. For example, you may have a certain model of how synapses work within the brain. There are certainly many different nonlinear effects involved there. The picture we saw earlier of the little vessel, which is separate from the membrane merging into it and then opening out into the intercellular space – all kinds of nonlinearities there.

There are many different particular realizations of this general model that I am proposing. So this is an idea, which thus far I have described in terms of classical physics. But, of course, we are not dealing with a world of classical physics. We have to talk about molecules and atoms and so forth if we are going to speak of the brain, so we certainly have to deal with quantum physics.

What do we do, then, to deal with quantum physics? Well, the same basic model can be carried over into the domain of quantum physics, and at this point the presentation draws very close to the presentation given by Dr. Stapp.

The first point that is needed in discussing quantum mechanics is that you have to have a quantum mechanical ontology. The standard approach in quantum mechanics is to say: never mind what is there, we just do calculations and compare the results with experiment. And of course, there you do not even have matter, what to speak of mind. So it is hard to talk about the relationship between the two.

You need an ontology. I will describe very briefly how you carry over this picture into the Heisenberg ontology, outlined by Henry Stapp. I have another picture just to give a sort of general idea of that.

What we are going to do is replace this smooth continuous path in classical physics by something a little fuzzy. In quantum mechanics basically things become fuzzy. As I described in terms of this idea of the cat paradox, in quantum mechanics, if you just let things run according to the Schrödinger equation, they get very fuzzy.

So you have to edit the wave equation in quantum mechanics, and you do this by collapsing the wave equation. These vertical lines I have drawn here represent the successive collapses of the wave equation, of the wave function, which is in between the lines evolving according to the Schrödinger wave equation.

Basically, what you postulate is that there is an actual history of real events, in which you have successive collapses of the wave function. What is really there is the actual history of real events, and that defines a somewhat fuzzy path. Not sharp like the classical path, but a fuzzy path. Quantum mechanical interference effects allow these fuzzy paths to do many different things, such as represent molecules, molecular bonding, and so many different things.

You can see how you could go over from this basic picture that I described to you in terms of adding simply the rule of nonlinear optimization to the classical laws of physics, to doing a similar thing in the quantum mechanical case. Now I would propose in fact, that this is how one could resolve the question of free will in the model introduced by Dr. Stapp, because he was introducing the idea that each collapse corresponds to an actuality of feeling or perception. This is all related to Heisenberg's interpretation of the wave function and so forth.

But the question then is: Does that make you a stochastic automaton? Is it that you are just moving essentially at random like a computer program in which you put in a random number generator to determine decisions? Or can you actually act according to your will?

What I am proposing here, is that by adding this feature of the nonlinear optimization rule, which is a very small addition to the laws of physics, it is possible to have a model in which all the known physical phenomena still take place, but you can have a "will" of a non-physical conscious entity directing and influencing the course of behavior of matter. So that is the basic model that I wanted to introduce.

I will close by pointing out a very interesting thing, as it seems to me. And that is that this model has some automatic empirical consequences, which we can deduce from everything that I have said thus far. In fact, the model has the consequence that the kinds of effects reported just now by Robert Jahn could broadly be expected to occur. Now, I cannot even begin to address all the very subtle fine points in the different phenomena that he was describing: different signatures of different operators and so many different effects that are there. But at least on a broad level, you would expect the kind of thing that he is describing.

So let me explain why that is. Imagine once again this idea of nonlinear optimization. It is easiest to think of this in terms of the classical model. What you can do, as I showed in that spring diagram before, is imagine this classical history to be a kind of spring, or slinky spring-like thing, and you are deforming it or bending it by imposing these different restraints.

The nature of this path, if we visualize it as a kind of spring, is that in some places it is stiffer than in other places. Naturally it is going to bend more in the flexible places. The flexible places correspond to the areas where you have deterministic chaos as a strong element of the physical interaction. And the inflexible regions correspond to clock-like mechanistic behavior.

One thing about the mechanism of a clock is that there is no deterministic chaos there, otherwise the clock could not keep time. So some patterns of physical interaction are not very flexible, others are. If you perform the nonlinear optimization, the system is going to bend in the area where the flexibility is.

Let me apply that to this random event generator study, which was just described to us. Let us suppose that a person does will for this generator to give more high numbers than low numbers. He is willing a certain outcome. In the model I described, his desire represents high numbers, and what his brain actually reports is something else. And that is a constraint.

You want to bring those together, minimize them, bring them together. But it is hard just to make the brain jump to a different thing. That would be a big violation of physics. It is also hard to make the LED display on the machine jump to a different number. That would be another big violation.

Similarly, all the way back to practically all of the electronics in the machine – because this is a very deterministic machine that I saw. It is not quantum – the electronics are meant to behave deterministically. All the way back within the machine, everything is very rigid and inflexible, until you reach the specific phenomena, such as white noise, that generate the randomness. Now in the case of electronic white noise, there would be very tiny molecular Brownian motion type things, which are ultimately very quantum mechanical. So that is where the flexibility is.

What you would expect to happen then, given this model, is that there would be changes, microchanges, in the randomness there, which would give rise to effects corresponding to the will of the operator.

The basic point I want to make here, is that on first seeing the results reported by Robert Jahn and Brenda Dunne, it may seem that this is really weird. How could such a thing be? How could there actually be evidence strong enough, along *these* lines, to make us overthrow, or over turn, all our ordinary conceptions so that we can accept such evidence as valid? Isn't it better just to think that there has got to be something wrong, and forget about it?

What I would propose is that actually, if you look at it in the context of this model, which was actually devised on the basis of the study of the free will problem, you can see that it is not such an unusual thing for these kinds of phenomena to occur. They are not necessarily so anomalous.

My main point in introducing that idea, is to actually try to encourage research into these kinds of phenomena and research into the idea that there may actually be such a thing as actual interaction between the conscious self and physical reality. The ideas are not so bizarre. They do not require such a radical change in what we are accustomed to thinking of, that these ideas are scientifically unconscionable.

So, with that, I will conclude. [*applause*]

Question (all questions read by Thompson): Well, the first question I have is, "What if you started with the premise that consciousness was holistic by nature? How would this fit or interfere with your trans-temporal approach to mind-brain interaction?"

Response: That is an interesting question. Saying consciousness is holistic by nature, one is referring there, I believe, to the idea that there is a kind of global consciousness.

It is sort of almost customary in these highly scientific discussions not to mention words like "God," but one may as well notice that that is what is being referred to: a global all-pervasive consciousness. So how would that fit into this approach I am presenting here, or how would that interfere with it?

I would propose that that fits in perfectly well. What I have been discussing here is the idea of the interaction between the individual conscious self, which we are thinking of in a dualistic sense here, and the brain of that person. Of course there are many people, and that means there are many individual conscious selves. But there can also be a universal conscious self. That is also possible. And that forms the subject matter of religion, theology, metaphysics, and so forth.

One way to look at this whole presentation is that it allows a place, or a relationship, to exist between hard science – physical science – and religion, in which the two are not necessarily seen as mutually exclusive. So that is one kind of consequence for these ideas.

You might ask, "Well, what about the will of the holistic consciousness? Could that influence matter?" Well, perhaps it could. That, of course, is a whole subject to go into, in theology and metaphysics.

Question: Here is another question: "How can the will caused by a brain function cause a change to the same brain function?"

Response: The idea here is that will, as I am presenting it in this model, is part of something distinct from the brain.

Now I can see that what we are willing at any particular time, indeed, has an awful lot to do with what is going on in our brain. For example, if I am thirsty, then I "will" to go and drink water. Now how did I will that? Well, my body was indicating thirst. That was reported to the brain through nerve impulses, and according to this idea, in addition to this process of volition that I have been talking about, you also have to have another process which I have not talked about. Namely, that the conscious self has to pick up information from the brain so as to find out what the state of the brain is.

So according to that idea, the conscious self would have to pick up that information, and translate that into its desire. That would be the basic concept in the dualistic picture. It is very interesting to go into the whole question of what is involved in this process of perception, but I cannot begin to go into that aspect right at the moment.

[*five minute break in recording*]

Question: "How could microscopic information of an orderly nature be introduced into a physical system by its unfolding from an enfolded implicate order."

Response: This is clearly a reference to David Bohm's "implicate order" idea.

I once asked David Bohm if he thought that orderly information for all kinds of different things that are going to happen is basically bound up in the implicate order there, and then is unfolding. As far as I could understand it, he did think that that was the case.

So that is an interesting idea. Of course, what is the "implicate order"? I think you will have to ask David Bohm to really explain that in detail. But I imagine that he would, in fact, believe that this could happen, and that is part of his concept, his implicate order model.

It is somewhat different from the present model though. I would like to emphasize that one thing I am proposing here when I talk about a dualistic picture of mind and matter, and that is just as matter is variegated – there are all kinds of different patterns and arrangements of matter in nature – so I would also propose that mind and consciousness are variegated. There are all kinds of different minds. We all have minds. We have already spoken of the idea of the universal mind. And there could be many other possibilities also, which we may not even know about.

So, I do believe that David Bohm's concept is monistic, to as great a degree as he can make it so.

Question: "Given that supposed predecessors often have no understanding of the physical basis of the phenomenon which they are trying to influence, how could they know which functions to optimize, i.e., which physical parameters to consider as targets for micro-influence."

Response: Well, of course, that was precisely the question that I was trying to answer with this model. That is the question that arises given the first model that I proposed, namely, that if micro-influences are generated by the conscious self, how then does the conscious self know how to influence micro events somewhere in the brain?

The conscious self does not even know anything about the brain, what to speak of micro events in a random number generator in which someone does not even know what the circuitry in that machine is. It is just a box, as far as they know.

How could you possibly know that? You could say: Well, subconsciously you do it. But that is really begging the question. So, the specific point of this model was that by introducing this simple, non-linear optimization rule, then you can solve that problem.

Question: "Does free will violate the second principle of thermodynamics?"

Response: Yes, I said that.

Question (continued): "If so, how does it get away with it?" [*laughter in audience*]

Response: Well, it should be ashamed of doing that!

Well, I guess I discussed how it happens, and that would be how it gets away with it.

Question: "What is the mind that "wills"? Is it matter?"

Response: Well, of course, this has been an extended discussion of the idea of mind-body dualism, which is one possibility. As I was saying in the beginning, we are considering alternatives here.

One may have a conviction that there is no mind-body dualism, and in fact that the solution to the free will problem that I gave right at the beginning of the lecture, is the right solution. And that is it.

But if you consider that it may be that there is mind-body dualism – and I do not think very many of us would be so bold as to say: I definitely know the answer to that. If they do say that, we would like to ask them how they know. Since it is a possibility, we are considering the idea that there is an actual fundamental difference between mind and matter.

Let me introduce the idea of a primitive element in a theory. In every theory that we have in science, there are irreducible elements. The idea of reductionism is that you reduce things down to your irreducible elements, and that is where you stop.

In the case, say, of old-time physics, you had electrons, protons, and neutrons. Those are irreducible elements, so you did not ask what a proton was. What is a proton made of? Well, a proton is made of a proton. It is a fundamental, irreducible element. Now, a later theory may come along that will explain a proton in terms of quarks. That is another thing.

So basically here, this mind or consciousness is an irreducible element. Although a later theory may come along, much more sophisticated, which talks about differentiation within consciousness. But that would be another whole topic.

Question: "Consciousness is not the same thing as 'will,' although they are related in free will. Or else do you believe that every mental event, i.e., perception or memory, is an act of 'will'?"

Response: Yes, this is a correct point. I have been emphasizing "will" here. The whole model that I am talking about has to do with will, volition, and so forth.

As I mentioned, there is also perception, and there are internal states of consciousness. There are feelings, all kinds of different states of perception. Our consciousness is a very complex thing. All complexity that we know of comes to us through our consciousness, so our consciousness is at least as complex as anything we *can* know of. So, I do not say consciousness is just will. That is just one aspect of consciousness.

Question: "Even within deterministic chaos, how does nature accommodate or yield to volition, even to a minute degree? Is nature somehow conscious of the 'will' on the part of the system which includes itself?"

Response: This is a good question. It brings up a very good point.

The point really is: Well, how does it do it anyway?

So, let me make a basic point about that. First of all, from the point of view of physics we never know how it does it, ever. All we ever do is write an equation, in which there are terms, mathematically.

For example, how does an electromagnetic field exert an influence on an electron? You will not learn that in physics. You will see a formula, and by the formula you calculate what happens with electrons and magnetic fields. But no one can really say *how* an electromagnetic field influences an electron. You merely postulate that it does, and that it does it according to a certain rule.

So, I am merely making a similar postulate here with regard to volition and the different physical parameters.

Now, I should then make another point about models. I conceive of models as being a very provisional thing. You make a model in order to deal with certain data. But you should not become attached to the model. And you should not try to reify the model and say: Well, this must be the absolute truth, this particular model.

You can consider, for example, the Bohr hydrogen atom model. Now, that was a very useful model. It explained a lot about atomic spectra, but look at how that has given way to the whole science of quantum mechanics.

So, I would conceive of a model like this as being a very provisional idea. In fact, I assume that it is wrong in some respects. At least it is very incomplete. But the idea is to take a start towards being able to actually talk about mind-body dualism, within the framework of science.

Question: "How does your theory differ from quantum cyclotron theory?"

Response: The difference between this theory, and theories of that sort – and first let me just briefly indicate that Professor Pribram has mentioned a number of different kinds of theories for what may be going on in the brain. You may have some kind of cyclotron propagating from place to place within the brain. That is a certain kind of waveform. You may have coherent interactions within the brain.. You may have some actual quantum phenomena similar to superconductivity.

All kinds of ideas may be there. These ideas are all well and good. The thing that I am presenting here is a completely different thing from that.

In all of those ideas, you are talking about cause and effect phenomena within the material system. There are many different kinds of cause and effect phenomena that you can deal with. What I am doing is taking the whole system with its cause and effect phenomena, whatever those may be – and we can study the details; they may be cyclotrons, they may be this or that – and adding this global non-linear optimization idea. So in that sense, it differs.

FIRST INTERNATIONAL CONFERENCE

on the study of

CONSCIOUSNESS WITHIN SCIENCE

February 17-18, 1990 San Francisco

The conference will examine the need for, and the methodological problems in the study of the phenomenon of consciousness from the perspective of a wide range of scientific fields.

INVITED PRESENTATIONS BY:

JOHN ECCLES *New concepts on the Mind-Brain Problem;* **KARL PRIBRAM** *Brain States and Processes as Determinants of the Contents of Consciousness;* **ROBERT JAHN** *Wave Mechanics of Consciousness;* **A.G. CAIRNS-SMITH** *Evolution and Consciousness;* **R.L. THOMPSON** *A Trans-Temporal Approach to Mind-Brain Interaction;* **E.C.G. SUDARSHAN** *Patterns in the Universe;* **JOHN SEARLE** *What is Wrong with the Philosophy of Mind Today?;* **HENRY STAPP** *A Quantum Theory of Consciousness;* **ROBERT ROSEN** *Consciousness: Immanent or Transcendent?;* **BRENDA DUNNE** *Engineering Anomalies Research;* **H. FROHLICH** *Biological Coherence;* **DAVID LONG** *Concepts of Consciousness: A Philosophical Critique;* **VASILII NALIMOV** *Spontaneity of Consciousness: A Probabilistic Theory.*

The program will also include question/answer sessions, panel discussions and a poster session. The format aims at affording registered participants ample opportunity for interaction with the distinguished guests.

REGISTRATION FEE: $125.00 before December 31, 1989; $150.00 thereafter. Full time students: $75/$85 (limited seats only). To register, please send check/money order made payable to the Bhaktivedanta Institute, together with personal details.

POSTER SESSION: Full paper/extended abstract (3 copies) due by December 31, 1989.

Please address all registration requests, paper submissions and inquiries to:

Ravi V. Gomatam, Organizing Secretary
THE BHAKTIVEDANTA INSTITUTE
84 Carl Street, San Francisco, CA 94117
Tel: 415-753-8647/8648
e-mail: bvi@cca.ucsf.edu

Poster announcing the 1990 BI conference in San Francisco.

Thompson (far right) along with John Searle (center) and Karl Pribram (left of Searle) in a discussion with the speaker.

Formal photo of conference participants.

CHAPTER 10B

A Trans-temporal Approach to the Laws of Physics

by Richard Thompson
La Jolla Institute, La Jolla, CA (c. 1986)

Notes:

The following paper (10b), presented in the form of a facsimile reprint, offers a formal development of many of the arguments presented in the previous presentation (10a).

Another manuscript located in the Archives with the title, "A Path-integral Approach to the Quantum Mechanical Measurement Problem," appears to be an earlier draft with a slightly different abstract.

At the end of (10b), there is an image of a letter written by Brian Josephson to Richard Thompson dated February 12, 1986, in which Josephson offers review commentary on this paper. Thompson had been working under Josephson beginning in August 1983 as a post-doctoral fellow at the Cavendish Laboratory, University of Cambridge, investigating quantum theory and theoretical biology. Of note, Josephson won the Nobel Prize in Physics 1973 "for his theoretical predictions of the properties of a supercurrent through a tunnel barrier, in particular those phenomena which are generally known as the Josephson effects" (www.nobelprize.org/prizes/physics/1973/josephson/facts/).

A TRANS-TEMPORAL APPROACH TO THE LAWS OF PHYSICS

by

Richard L. Thompson

La Jolla Institute
Division of Applied Nonlinear Problems
3252 Holiday Ct., Suite 208,
La Jolla, CA 92037

ABSTRACT

We present a basic formulation of classical and quantum mechanics in which these two theories take on a similar form. In this formulation, classical mechanics becomes a non-deterministic theory, and quantum mechanics becomes capable of modeling the observable physical world as an objective reality. We use findings from the study of deterministic chaos to illustrate the extreme indeterminism that this formulation of classical mechanics entails for some systems, and we also show how the quantum theory displays corresponding indeterminism when applied to these systems.

Statistical laws are introduced in a uniform way into both the classical and the quantum mechanical models. We show that by making simple modifications of these laws, it is possible for both models to predict teleological phenomena of the kind observed in some parapsychological studies. We also show that both models entail non-local phenomena similar to the EPR effect, and we show that the models are consistent with the special theory of relativity.

1. INTRODUCTION

The rise of classical mechanics marked the culmination of a major shift in Western thinking from the Aristotelian conception of the world as an organism to the conception of the world as a clocklike mechanism operating according to mechanistic laws.[1] Classical mechanics is based on equations of motion which are deterministic; given the exact position and momentum coordinates of a classical system at time t_0, it is possible in principle to calculate the exact state of the system at any time in the future or the past. This property has many profound philosophical implications. For example, if nature is indeed strictly governed by such deterministic laws, then sentient beings must be pure machines, devoid of free will (or else "free will" must be defined as the deterministic interaction of particles in the brain).

With the later development of quantum mechanics, physics ceased to be strictly deterministic, and many physicists have tried to reintroduce the idea of conscious volition into our physical world view.[2-5] However, the advent of quantum mechanics did more than simply stress the role of the observer and add an element of indeterminism. In its standard formulation, quantum mechanics requires us to renounce the idea of forming a coherent theoretical picture of objective reality. Rather, it enjoins us to see physical theories merely as computational systems designed to predict patterns of statistical correlation in observed data.[6,7]

In this paper we present a reformulation of both quantum mechanics and classical mechanics which brings the two theories closer together, adding an element of indeterminism to classical mechanics, and also showing how quantum mechanics can be seen to provide an objective model of the world of our experience. This reformulation is based on the idea that the laws of physics should be seen as global statements about the history of events in space-time as a whole, rather than as relations of cause and effect governing the unfolding of events with the passage of time. Its starting point is Hamilton's principle in classical mechanics and the Feynman path integral in quantum mechanics, both of which present the laws of physics from a global viewpoint.

Using these starting points, we build up the idea of a process of trans-temporal selection which picks out a particular history of macroscopic events in space-time that conforms with the classical or quantum mechanical laws of nature. This process generates a specific space-time history which can be regarded as a representation of objective reality. However, although the process acts in accordance with the physical laws, it is not rigidly determined by them, and thus the laws do not precisely specify this history.

In Section 2 we review Hamilton's principle and introduce the process of trans-temporal selection for classical physics. In the next section we draw from the theory of deterministic chaos to illustrate the high degree of indeterminism allowed by our reformulation of classical mechanics. In Section 4 we show how statistical laws can be incorporated into this fomulation, and we discuss the definition of the arrow of time.

In Section 5 we discuss the various interpretations that have been given to the formalism of quantum mechanics. This is followed in the next two sections by the introduction of the quantum mechanical form of the process of trans-temporal selection. In Section 8 we argue that this process is compatible with special relativity, and we discuss the EPR effect and the idea of nonlocal interactions. In Section 9 we argue that the process of trans-temporal selection leads to certain predictions which seem to be confirmed by some of the data gathered in recent years in studies of psychokinesis. Finally, in Section 10 we make some concluding remarks and mention some avenues of future research.

2. CLASSICAL MECHANICS

We will begin our discussion of classical mechanics by defining Hamilton's principle for a simple classical system consisting of n particles of mass m with coordinates $\bar{Q} = (Q^1,...,Q^{3n})$. Consider an arbitrary path Q(t) defining the history of the system from time $t=t_a$ to time $t=t_b$, and constrained by the requirement that $\bar{Q}(t_a)=\bar{Q}_a$ and $\bar{Q}(t_b)=\bar{Q}_b$. We can define a quantity called the action which is a global property of this entire history of events:

$$A(\bar{Q}) = \int_{t_a}^{t_b} L(\bar{Q}(t),\dot{\bar{Q}}(t),t)\, dt \qquad (1)$$

where the Lagrangian $L(\bar{Q},\dot{\bar{Q}},t)$ expresses the laws of physics for the system. The action A will generally vary as we vary the events in our arbitrary path. But if the action is invariant under all infinitesimal changes in the path between times t_a and t_b, then it turns out that this path satisfies the classical equations of motion.

We can define the sensitivity of the action to variations in the path as

$$S(\bar{Q}) = \int_{t_a}^{t_b} \left[\sum_{i=1}^{3n} |\delta A(\bar{Q})/\delta Q^i(t)|^2 \right]^{1/2} dt \qquad (2)$$

This quantity is simply a measure of how closely the path follows the classical equations of motion. Thus, for a system with $L=m\dot{\bar{Q}}\cdot\dot{\bar{Q}}/2-V(Q)$ the sensitivity is

$$S(\bar{Q}) = \int_{t_a}^{t_b} dt\, |m\ddot{\bar{Q}}(t)+\nabla V(\bar{Q}(t))| \qquad (3)$$

In the usual formulation of classical mechanics, only paths from $\bar{Q}_a$ to $\bar{Q}_b$ for which the sensitivity S=0 are regarded as having physical significance. We might ask, however, whether or not paths with S nearly equal to zero might also be significant. It turns out that for a broad set of classical systems there is a very large set of paths for which the sensitivity is very close to zero. Indeed, many of these paths differ greatly from the classical solution, but come so close to following the classical laws of motion that no experimental measurements occurring within the system could show any deviation from these laws.

This suggests a new way of looking at the classical laws of motion. We can regard them as defining a large class of paths from $\bar{Q}_a$ to $\bar{Q}_b$ which come so close to satisfying these laws that we could not hope to measure their deviation from them. We can call this the set of near-classical paths, and define it as

$$NC(\epsilon) = \{\text{Paths } \bar{Q} \text{ from } (t_a,\bar{Q}_a) \text{ to } (t_b,\bar{Q}_b) \mid S(\bar{Q})<\epsilon\} \quad (4)$$

where $\epsilon>0$ is very small. We postulate that the classical laws require the path followed in nature to lie in the set $NC(\epsilon)$, but that they do not specify which particular member of $NC(\epsilon)$ this path will be.

We can imagine that this path is selected in nature by an agency which conforms with the laws, but nonetheless is not rigidly constrained by them. We will refer to the action of this agency as the process of trans-temporal selection. We argue in later sections that various statistical regularities observed in nature can be accounted for by means of a probabilistic definition of this process. In this paper, however, we will not try to specify the exact nature of the agency that lies behind the selection process.

3. DETERMINISTIC CHAOS

In order to see the significance of this way of looking at the classical physical laws, it is necessary for us to get some idea of the nature of the near-classical paths. We can do this by examining some of the results found recently in the investigation of what is called deterministic chaos.

The term deterministic chaos refers to the observation that many classical systems respond with such sensitivity to small variations in position and momentum that their future behavior is chaotic and unpredictable in practice, even though it is exactly predictable in principle. Suppose $\bar{Q}$ is a path that exactly satisfies the equations of motion. Suppose that $\bar{Q}'$ is a path obtained by following $\bar{Q}$ up to time t, making very slight and gradual changes in some of the system's position and momentum variables over the interval from t to $t+\Delta t$, and letting the system evolve according to the equations of motion after $t+\Delta t$. Then, $\bar{Q}'$ nearly satisfies the laws of motion (so that it is a near-classical path), and it agrees with $\bar{Q}$ initially. But it may be that $\bar{Q}'$ greatly differs from $\bar{Q}$ shortly after $t+\Delta t$.

We can get a quantitative idea of the meaning of the terms "nearly" and "greatly" by considering some specific examples. We will begin by discussing a simple example of a chaotic system that was introduced and extensively studied by Henon.[8]
Consider the two-dimensional potential

$$V(x,y) = (x^2+y^2)/2 + x^2y - y^3/3 \quad (5)$$

Henon and his coworkers have shown that the classical motion of a particle in this potential can be very hard to predict when the energy of the particle is above a certain level. This is illustrated in Fig. 1 for a particle of mass m=1 gram and E=1/6 (using cgs units). The two-dimensional region in the figure represents a "surface of section" through the four-dimensional phase space of the system, and the many isolated points represent passages of a single particle trajectory through this surface.

As time passes, these points jump around on the surface of section in an apparently random fashion, and it appears that they will fill up a broad area in the limit as time goes to infinity. (In contrast, the points will remain confined to a single smooth curve in a "well behaved" classical system.)

This apparently random behavior of the particle's deterministic trajectory is evidently due to the fact that extremely small variations in the position or momentum of the particle will tend to be quickly amplified to a large degree. In fact, this amplification is so great that variations in position and momentum of the kind specified by the Heisenberg uncertainty principle can quickly give rise to macroscopic effects, even in the case of a particle of mass m=1 gram. To show this we have calculated a measure of amplification for a particular trajectory.

Let $\bar{x} = (p,q)$ represent a point in the 4-dimensional phase space of the system, and let $T_t(\bar{x})$ designate the phase space position at time t of a particle of mass m=1 that started at $\bar{x}$ at time t=0, and moved according to the Henon potential. If $\bar{\xi}$ is very small, then

$$T_t(\bar{x}+\bar{\xi}) - T_t(\bar{x}) \approx M(t,\bar{x})\bar{\xi} \tag{6}$$

where $M(t,\bar{x})$ is the Jacobian matrix of the transformation T_t at $\bar{x}$. The magnitudes of the entries in $M(t,\bar{x})$ thus provide a measure of the deviation of $T_t(\bar{x}+\bar{\xi})$ from $T_t(\bar{x})$ caused by a small variation $\bar{\xi}$ in $\bar{x}$.

Fig. 2 depicts a closed orbit of period τ=34.86 sec for which $M(\tau,\bar{x}_0)$ can be readily calculated for a starting point $\bar{x}_0$. $M(\tau,\bar{x}_0)$ contains entries with magnitudes as large as 10^3, and its 4th and 8th powers contain entries with magnitudes of 4.2×10^{12} and 2.7×10^{26}. Suppose that the path $\bar{Q}$ is taken to be this orbit for the 1 gram particle, and that a variation in position or momentum of order 3.7×10^{-27} is made in the interval from t to t+Δt. If the variation is directed properly, the corresponding path $\bar{Q}'$ will deviate substantially from $\bar{Q}$ after a time of 8τ. For still later times the path $\bar{Q}'$ will not even resemble the closed orbit $\bar{Q}$.

According to the Heisenberg uncertainty principle, two conjugate position and momentum coordinates must have inherent uncertainties Δq and Δp with $\Delta q \Delta p \geq \hbar/2 \approx 5\times10^{-28}$. Here, however, we find that we need to know positions and momenta with greater accuracy than this to determine the classical trajectory over relatively short periods of time. Thus this system is so sensitive to small changes in initial conditions that to analyze it we must take quantum mechanical uncertainty into account.

In a system with extremely sensitive dependence on initial conditions, it is possible for the classical path followed by the system to take on a great variety of complex forms. To indicate how great this variety can be, let us consider a simple example based on an old fashioned pinball machine.

Consider an infinite array of fixed disks of radius r, centered on the lattice points (x,y) of the plane. Assume that another disk of radius r is moving in the area between these disks without friction, and is bouncing from disk to disk with elastic collisions. If we choose a small enough r ($r \leq 1/40$ will do) then we can show the following:

Theorem 1. Let B = {(0,1),(1,1),...} be the set of 8 lattice points immediately surrounding the origin (0,0). Let $\{\bar{z}_1,...,\bar{z}_n\}$ be any sequence of lattice points which

(1) proceeds by single steps (so that $\bar{z}_k-\bar{z}_{k-1} \in B$ for $k=2,...,n$), and

(2) never proceeds in the same direction for two steps in a row (so that $\bar{z}_k-\bar{z}_{k-1} \neq \bar{z}_{k+1}-\bar{z}_k$ for $k=2,...,n-1$.)

Then we can choose a starting point $\bar{x}_0$ and an initial angle of motion θ_0 so that $\{\bar{z}_1,...,\bar{z}_n\}$ specifies the sequence of fixed disks encountered by the moving disk in its first n bounces.

This theorem is proven in Appendix 1. It shows that the initial angle θ_0 of the moving disk can be chosen so that the disk follows any one of a wide variety of desired paths, even though it also moves strictly in accordance with the classical deterministic laws for frictionless motion and elastic collision. Starting with $\bar{z}_1=(0,0)$, there are 7^{n-1} paths $\bar{z}_1,\ldots,\bar{z}_n$ satisfying conditions (1) and (2) of the theorem. The overwhelming majority of these paths will appear to be completely random in form, but paths can also be specified that express any desired string of coded information. For example, the path $\bar{z}_1,\ldots,\bar{z}_n$ could be chosen so that it spells out Shakespeare's plays in a particular style of handwriting.

Whether the path that is generated is orderly or chaotic, we can look at it as a step by step unfolding in time of information that is stored up in the initial angle θ_0. Indeed, to specify a typical n-step path out of 7^{n-1} paths, we need at least $(n-1)\log_2(7)$ bits of information, and thus we need to specify at least this many bits in the binary expansion of θ_0. We can think of the bouncing disk as steadily reading out and expressing information that is stored up in higher and higher order digits in the binary expansion of θ_0.

We can also look at the path of the disk from the trans-temporal viewpoint adopted here. From this viewpoint, the classical laws of nature are regarded as rules telling how events adjacent in time are connected together. The inherent flexibility of these laws is sufficient to enable the pattern in events unfolding over large time periods to be freely chosen, even though the system seems to evolve by the classical laws over short time periods. In this example, it turns out that the classical laws can be exactly satisfied at all times, even though the large-scale pattern can be chosen with great freedom. In general, however, we require only that the classical laws should be followed closely enough so that they are never violated to an observable degree.

When the disk's path is looked at in this way, the value of θ_0 no longer appears to be as important as it does when it is viewed as the starting point of a chain of causes and effects which completely determines the future of the system. Now the value of θ_0 arises simply as a consequence of the over-all selection of the path. Indeed, if we do not require the classical laws to be strictly obeyed at all times, then the path will not have to be so precisely defined that it forces the binary expansion of θ_0 to encode information for large numbers of successive $\hat{z}_k$'s. In this case only a few digits of this expansion will be significant.

Thus far we have given examples only of systems with a small number of degrees of freedom. However, the ideas that we have illustrated can also be applied to highly complex systems. We suggest that the trajectory followed by a general classical system of n bodies can be viewed mathematically as a flexible continuum which can be bent into a large variety of complex shapes for which the classical laws of motion are nearly obeyed at all times. In situations in which the classical trajectory is highly predictable, this process of selection can produce only trajectories that are very close to exact classical solutions with S=0. This will be true, for example, for exactly integrable systems. However, in situations amenable to deterministic chaos the process of selection will allow for nearly lawful trajectories of highly arbitrary complex form.

The phenomenon of turbulence in fluid flow provides an example of a complex system that appears to exhibit deterministic chaos. Empirical and numerical studies suggest that the large scale motion of the vortices in turbulent flow is strongly effected by the amplification of extremely small variations in the current, and thus the behavior of these vortices is inherently unpredictable.[9] In this situation $NC(\epsilon)$ should contain paths representing many different patterns of turbulent flow, and the process of trans-temporal selection can be viewed as randomly picking out one of these paths.

4. STATISTICAL LAWS

In principle, the process of trans-temporal selection can generate paths in which highly ordered structures seem to arise spontaneously from a state of initial disorder. Yet we normally do not expect to see this, and we expect instead to observe the appearance of a great deal of apparently random noise. This tendency for disorder to increase at the expense of order defines the direction of the passage of time, and as is well known, it is not a consequence of the time reversible laws of motion.

In mathematical models of physical systems this tendency is normally introduced by adding probabilistic laws to the original laws of motion. For example, this is done in statistical mechanics by defining the initial conditions of the system by means of probability distributions known as thermodynamic ensembles. In quantum mechanics we also see that the element of randomness has to be added to the deterministic equation of motion (the Schrödinger equation) as an independent statistical law (von Neumann's postulate of wave function collapse).[10]

The simplest way of introducing thermodynamic randomness into the process of trans-temporal selection is simply to postulate that this process chooses a path "at random" with a probability density proportional to

$$\mu(\bar{Q}) = F(\bar{Q})T[S(\bar{Q})<\epsilon] \qquad (7)$$

where T[condition] is 1 if the condition is true and 0 if it is false.

The function $F(\bar{Q})>0$ places a limitation on the possible paths. For example, if F restricts the paths at times near some t_* in the remote past to a certain highly ordered state of affairs, then trans-temporal selection according to Eqn. (7) should produce a path in which this situation gradually deteriorates from that time onward, and time's arrow goes from past to future. In contrast, if F restricts the paths at times near some t_* in the remote future to a highly ordered state (and imposes no restrictions at other times), then the selection process should produce a path for which time's arrow proceeds from future to past. Of course, we normally expect the first of these two situations to prevail. However, there is some empirical evidence suggesting a significant role for more complex F's, which impose restrictions on the paths at various times. This is discussed in Section 9.

5. QUANTUM MECHANICS

The history of the quantum theory has been marked by a chronic controversy over what is generally known as the quantum mechanical measurement problem. As we shall see, the trans-temporal approach to the laws of physics provides a new way of tackling this problem. We will therefore give a brief description of some of the fundamental issues involved in this controversy and mention some of the main approaches that have been taken to resolve it.

Broadly speaking, there are two ways in which a mathematical theory can describe nature. These are:

(1) The theory quantitatively predicts the patterns of correlation in experimental measurements.

(2) The theory contains mathematical structures which (in principle) represent "objective reality" in a one-to-one fashion.

Here "objective reality" may refer to the totality of what exists in nature, or it may refer to certain limited aspects of nature that a particular theory is intended to describe.

Since acts of measurement are part of reality, a theory satisfying (2) must also satisfy (1). Up to the close of the nineteenth century, it was generally taken for granted that a well developed physical theory should satisfy both (1) and (2). However, the striking feature of quantum mechanics is that while it satisfies (1), it has proven very difficult to find a formulation of the theory which could satisfy criterion (2).

The reason for this is that the state vectors (or wave functions) used to describe nature in quantum mechanics are capable of simultaneously describing many macroscopically distinct states of affairs. There are many situations in which a state vector evolving in accordance with the Schrodinger equation will naturally develop such macroscopic ambiguities. The most famous example is Schrödinger's cat paradox, in which the state vector comes to describe a cat which is simultaneously dead and alive. In quantum mechanics the state vector provides the most complete description of the system that is possible, and thus if such macroscopically ambiguous state vectors are allowed, then it follows that the theory does not give a one-to-one representation of physical reality as we normally conceive of it.

The standard approach to quantum mechanics is to simply accept this state of affairs and adopt (1) as the ultimate criterion for a successful physical theory. According to this approach, it is inherently impossible for a physical theory to provide an intelligible picture of the reality underlying experimental observations. Wigner[6] calls this "the most natural epistemology of quantum mechanics," and it is hard to find fault with it logically or empirically.

However, many attempts have been made to modify or reinterpret quantum mechanics in such a way as to obtain a theory satisfying criterion (2). These include:

(1) Theories involving the automatic "collapse" of the wave function. Here the idea is that as the wave function spreads out through configuration space, it is repeatedly replaced by a modified wave function that is not macroscopically ambiguous. This mathematical procedure of collapse is thought to correspond to an actual process occuring in nature. Some authors have thought that this process is implicit in the formalism of quantum mechanics as it stands,[11] but others have argued that this is not so.[12] A number of physicists have proposed that the collapse of the state vector is connected with the making of an observation by a conscious observer,[2-4] and Wigner in particular has suggested that the laws of quantum mechanics as we now know them will have to be completely reformulated in order to take into account the existence of consciousness as a real feature of nature.

(2) "Hidden variables" theories. Here new parameters are added to the state vector formalism of quantum mechanics to bring it into one-to-one correspondence with reality as we normally conceive of it. A prominent example is the "quantum potential"

theory of David Bohm[13]; a very similar theory is provided by the stochastic mechanics of Edward Nelson.[14]

(3) The "many worlds" (or EWG) theory of Everett, Wheeler, Graham.[15] Here a state vector is posited for the universe as a whole, and this state vector is taken as a direct, one-to-one representation of objective reality. The many macroscopically distinct components of this state vector are interpreted as distinct universes which exert no measurable influence on one another, and thus our idea of objective reality is extended in such a way as to conform with the quantum mechanical formalism.

In the next section we will show that the trans-temporal approach to the laws of physics can be used to develop an additional formulation of quantum mechanics which satisfies criterion (2) for a physical theory.

6. A TRANS-TEMPORAL FORMULATION OF QUANTUM MECHANICS

Our point of departure for reformulating the quantum theory is provided by the Feynman path integral. In the usual formulation of quantum mechanics, the state vector (or wave function) can be thought of as a wave propagating through a multi-dimensional space of parameters. Speaking loosely, the intensity of this wave at one point in space and time can be expressed as a sum of effects derived from nearby points on the wave at a slightly earlier time. Each of these effects can in turn be expressed as a sum of effects deriving from nearby points at a still earlier time, and so on for earlier and earlier times. (This is essentially Huygens' principle.) Thus we can view the wave at one space-time point as being a sum of contributions made along the many different paths leading up to that point.

Feynman[16,17] expresses this idea through the following formula, which he calls a path integral:

$$K_C(t_b,\bar{z}_b;t_a,\bar{z}_a) = \int_C \exp[iA(\bar{z})/\hbar]\ D\bar{z} \qquad (8)$$

Here t_a and t_b are two successive times, and $\bar{z}_a$ and $\bar{z}_b$ are two values for the configuration of the system at those times. The parameter z represents a function $\bar{z}(t)$ defining a path. $A(\bar{z})$ is the action for the path $\bar{z}$, as defined in Eqn. (1), and C is a set of paths from $(t_a,\bar{z}_a)$ to $(t_b,\bar{z}_b)$ defining the range of the integration. If C is the set of all paths spanning these limits, the integral can be viewed as the value of a quantum wave function at time t_b and location $\bar{z}_b$, given that the wave function evolved according to the Schrodinger equation corresponding to the action A, and that it was concentrated at $\bar{z}_a$ at time t_a. (We note that it is difficult to define $D\bar{z}$ as a mathematical measure, but rigorous mathematical definitions of the path integral have been made. For example, see DeWitt.)[18]

For some paths $\bar{z}$, the action $A(\bar{z})$ will be sensitive to small variations in $\bar{z}$, and thus the sum of terms $\exp(iA(\bar{z})/\hbar)$ over these nearby paths will tend to cancel out. For other paths, this sensitivity will be less, and the corresponding sum over nearby paths will take on a non-zero value. Thus we can see that the path integral will tend to give weight to paths that approximate the criterion of stationary action that is used to define the classical trajectory of a system. Indeed, the difference between classical and quantum mechanics can be seen as the consequence of allowing many nearly stationary paths, rather than a single exactly stationary path.

Our reformulation of quantum mechanics is based on the following idea: Let us suppose that a physical system actually follows a fuzzy trajectory, called a channel, consisting of a set of similar paths. Feynman argues that the (unnormalized) probability density that the path followed by a quantum mechanical system lies in such a channel should be given by $|K_C(t_b,\bar{z}_b;t_a,\bar{z}_a)|^2$, where the C in Eqn. (8) is now confined to the set of paths constituting this channel.[17] Let us therefore suppose that nature somehow selects a channel C, and that this is done at random in accordance with the probability density given by $|K_C|^2$.

We will refer to the selection of a particular channel in nature as the process of trans-temporal selection. If this selection is performed in accordance with the probabilities $|K_C|^2$ defined by Eqn. (8), then the channel that is selected should represent a sequence of events that conforms with the statistical laws of quantum mechanics.

As we observed in the discussion of Eqn. (7), these

probabilities can also be modified by multiplying them by a function F(C) which picks out certain channels in preference to others. This gives us the following quantum mechanical version of Eqn. (7).

$$\mu(C) = F(C)|K_C(t_b,\bar{z}_b;t_a,\bar{z}_a)|^2 \qquad (9)$$

By introducing such an F, it is possible to define the arrow of time for the system in the same way that we did this in Section 4 for classical mechanics.

This formulation of quantum mechanics corresponds quite closely to our reformulation of classical mechanics. We can see this by applying it to a simple classical system with n degrees of freedom. Let a be a fixed positive number, and for $i=1,\ldots,n$ let $c_i(t)$ be a real function on the interval from t_a to t_b. The function $\bar{c}(t)=(c_1(t),\ldots,c_n(t))$ defines a path through the n-dimensional configuration space of the system, and we can define a channel C to be the set of all paths $\bar{z}(t)$ satisfying

$$c_i(t)-a \leq z_i(t) \leq c_i(t)+a \qquad (10)$$

for $t_a \leq t \leq t_b$ and $i=1,\ldots,n$. (For the path integral to be nonzero we require that the endpoints $\bar{z}_a$ and $\bar{z}_b$ of the paths $\bar{z}(t)$ must lie within the channel.)

Such a channel and one of its enclosed paths is shown in Fig. 3. For simplicity we shall deal with a system defined by the Lagrangian $L = m|\dot{\bar{z}}|^2/2 - V(\bar{z},t)$, where $V(\bar{z},t)$ is a (possibly) time dependent potential. We will denote the path integral in

Eqn. (8) as $K_{V,C}(\bar{z}_b;\bar{z}_a)$, where the parameters t_a and t_b have been left implicit, and the subscript V has been added to express dependence on the potential V.

$K_{V,C}(\bar{z}_b;\bar{z}_a)$ can be thought of as expressing the degree to which a wave function concentrated at $\bar{z}_a$ at time t_a is able to traverse the channel and reach $\bar{z}_b$ at time t_b. This depends on the potential V which specifies the physical interactions occurring in the system. For channels conforming to the path(s) that such a wave function will naturally tend to follow, $K_{V,C}(\bar{z}_b;\bar{z}_a)$ will take on a non-zero value, but for channels following completely different paths, it will be zero or nearly zero.

When we speak of a wave "traversing" the channel C, we mean that the wave satisfies the ordinary Schrodinger equation with potential V within the channel, and that the wave is reduced to zero along the boundary of the channel by an additional imaginary potential. Thus the path integral of Eqn. (8) corresponds to a wave function satisfying the equation

$$i\hbar\partial\Psi(\bar{z})/\partial t = (H + V_C(\bar{z},t))\Psi(z) \qquad (11)$$

where $V_C(\bar{z},t) = 0$ when $|z_i - c_i(t)| < a$ for $i=1,...,n$, and $V_C(\bar{z},t) = -i\infty$ otherwise.

We define the free propagation value for a channel of width 2a to be $K_{0,0}(\bar{\xi}_b;\bar{\xi}_a)$. This expression gives the degree to which a wave function concentrated at $\bar{\xi}_a$ at time t_a will be able to freely propagate ($V = 0$) through a straight channel ($c_i = 0$), and reach $\bar{\xi}_b$ at time t_b. If we disregard the possibility of a potential that focuses the wave function within a given channel, we can see that $K_{0,0}(\bar{\xi}_b;\bar{\xi}_a)$ represents the maximum level of wave transmission that we can expect through a channel of width 2a and time span t_b-t_a.

In general, in a quantum mechanical system it will not be possible for $|K_{V,C}(\bar{c}_b+\bar{\xi}_b;\bar{c}_a+\bar{\xi}_a)|$ to be as large as $|K_{0,0}(\bar{\xi}_b;\bar{\xi}_a)|$. This is due to the fact that the potential may cause the wave function to spread out widely, and thus be absorbed by the channel walls much more rapidly than will occur in the case of free propagation. Such spreading will occur, for example, in the situation of Schrödinger's cat, in which the wave function bifurcates into branches representing live and dead cats. The channel will be wide enough to accomodate only one of these alternatives, and the wave representing the other one will be absorbed.

In a system involving only macroscopic variables, however, the wave function may spread out at the rate of a freely propagating wave. The following theorem shows this for the case that we are considering:

Theorem 2. Let $a>0$, the channel C, and the potential V be as defined above. Define

$$s(\bar{c},t,\bar{y}) = |m\ddot{\bar{c}}(t)\cdot\bar{y} + V(\bar{c}(t)+\bar{y},t) - V(\bar{c}(t),t)| \qquad (12)$$

Let $\bar{c}(t_a)=\bar{c}_a$ and $\bar{c}(t_b)=\bar{c}_b$, and assume that $|\xi_{a,i}|\leq a$ for $i=1,\ldots,n$. Then

$$\left[\int_{[-a,a]^n} \left| |K_{V,C}(\bar{c}_b+\bar{\xi}_b;\bar{c}_a+\bar{\xi}_a)| - |K_{0,0}(\bar{\xi}_b;\bar{\xi}_a)| \right|^2 d\bar{\xi}_b \right]^{1/2}$$

$$\leq \exp\left[\int_{t_a}^{t_b} \sup_{\bar{y}} s(\bar{c},t,\bar{y})\, dt/\hbar \right] - 1 \qquad (13)$$

where the sup is over $\bar{y}$ for which $|y_i| \leq a$, $i=1,...,n$. This theorem is proven in Appendix 2. It indicates that if a is properly chosen, then a channel following a near-classical path of low sensitivity (given by Eqn. (3)) will conduct a wave function with no greater loss than occurs in a straight channel of width 2a carrying a freely propagating wave. Since we would not generally expect a lower level of loss than this (except perhaps in the case of special, focusing potentials), this means that near-classical paths will tend to be selected by our quantum mechanical process of trans-temporal selection. Since near-classical paths can take on a wide variety of shapes, this means that the paths generated by the quantum mechanical selection process for a "classical" system can also exhibit great variety.

We can illustrate Theorem 2 by applying it to the Henon potential described above. For this potential we can show by a few calculations that

$$\int_{t_a}^{t_b} \sup_{\bar{y}} s(\bar{c},t,\bar{y})dt \leq \sqrt{2}a \int_{t_a}^{t_b} |m\ddot{\bar{c}}(t)+\nabla V(\bar{c}(t))|dt$$

$$+ (t_b-t_a)a^2(3+\sqrt{3}+4a/3) \qquad (14)$$

where the sup is over $\bar{y}$ with $|y_i|\leq a$, $i=1,2$.

Here the first term on the right hand side is a measure of the deviation of the channel function c(t) from a classical path of the system, and the second term is a constant. The integral in the first term is a measure of how well $\bar{c}(t)$ obeys the classical equation of motion, $m\ddot{\bar{c}}(t)=-\nabla V(\bar{c})$. We have already shown that it is possible for $\bar{c}(t)$ to adhere closely to this equation, even though it deviates considerably from a classical trajectory (for which this term is 0).

Eqn. (14) can be used to estimate an appropriate value for the parameter a defining the width of the channel. If a is too large, then the rightmost term in Eqn. (14) will be large, and we won't be able to conclude that near-classical channels have high quantum mechanical probabilities. On the other hand, if a is too small, then the right hand side of Eqn. (14) will always be small, and all channels will have similar quantum mechanical probabilities (of nearly 0). These considerations lead to the following inequalities for a:

$$\hbar/\sqrt{2} << a << .45[\hbar/(t_b-t_a)]^{1/2} \qquad (15)$$

where we assume that the path remains within the triangular potential well shown in Fig. 1. For example, if $t_b - t_a$ is 10^{10} seconds, then a should lie between 10^{-27} and 10^{-19} cm.

7. COMPLEX QUANTUM SYSTEMS

Thus far we have considered channels involving functions $c_i(t)$ corresponding to position coordinates z_i for a number of particles. These systems are essentially classical and macroscopic in character. In a general complex system the channel should be defined using macroscopic variables defining positions, densities, or perhaps field strengths. Purely microscopic variables such as electron spin can be left unconstrained by the channel.

An example of a complex system is provided by an elastic medium satisfying the wave equation,

$$\mu\eta_{i,00} - K\eta_{i,jj} = 0, \qquad (16)$$

where $\mu=\mu(x,y,z)$ represents the density of the undisturbed medium as a function of position, and K is the medium's modulus of elasticity. Here we use the notation $(t,x,y,z) = (x^0,x^1,x^2,x^3)$, and we adopt the convention that double Latin indices are summed over 1,2,3. A subscript k preceeded by a comma indicates partial differentiation with respect to the corresponding x^k.

In this equation $\bar{\eta}(t,x,y,z)$ represents the displacement at time t of an infinitesimal part of the medium that has an equilibrium position of (x,y,z) and a mass of $\mu(x,y,z)d^3x$. The solutions of the equation represent macroscopic sound waves propagating through the medium. The action is given by

$$A = 1/2 \int d^4x[\mu\eta_{i,0}\eta_{i,0} - K\eta_{i,j}\eta_{i,j}] \tag{17}$$

and we can define channel functions $c_i(t,x,y,z)$ for i=1,2,3. The channel C corresponding to the c_i's is defined to be the set of functions $\eta_i(t,x,y,z)$ for which

$$\|\bar{\eta}(t)-\bar{c}(t)\|_2 \leq \epsilon \tag{18}$$

for all $t_a \leq t \leq t_b$. Here $\|\cdot\|_2$ is the L_2 norm over x,y,z. This channel consists of functions $\bar{\eta}(t)$ which are macroscopically similar to the channel function $\bar{c}(t)$, but which may differ from it considerably on the microscopic level. Thus the channel prescribes the macroscopic situation, but at the same time allows freedom for microscopic quantum mechanical fluctuations.

We define the path integral over this channel to be

$$K_{V,C}(\bar{\eta}_b;\bar{\eta}_a) = \int_C \exp(iA(\bar{\eta})/\hbar)\, D\bar{\eta} \tag{19}$$

where $\bar{\eta}$ stands for the functions $\eta_i(t,x,y,z)$, $\bar{\eta}_b$ stands for the functions $\eta_i(t_b,x,y,z)$ for t fixed at t_b, and $\bar{\eta}_a$ stands for the corresponding functions at $t=t_a$. $\bar{\eta}_a$ and

$\bar{\eta}_b$ should satisfy the channel condition of Eqn. (18) at $t = t_a$ and t_b. Here the potential (or potential density) V is defined to be

$$V = K\eta_{i,j}\eta_{i,j} \tag{20}$$

It can be shown that if $\mu(x,y,z)$ varies in a periodic fashion in the x, y, and z directions, then Eqn. (16) will have chaotic solutions. For example, suppose that $\mu(x,y,z)$ takes on high values in spherical regions of macroscopic radius R surrounding the positions (iA,jA,kA), where A>2R and i,j,k vary over the integers. Then, by analogy to the pinball problem, we would expect that small changes in a sound wave at time t will be greatly amplified over a relatively short time. This indicates that there should exist many macroscopically distinct sound wave patterns which are near-solutions of Eqn. (16). If we analyze this problem quantum mechanically using the trans-temporal approach, we can show that there are macroscopically distinct quantum mechanical solutions corresponding to these near-solutions of Eqn. (16). Thus, an elastic medium with a regular pattern of density inhomogenieties should exhibit macroscopic quantum mechanical indeterminism.

This is shown by the following theorem, which is analogous to theorem 2:

THEOREM 3. Given the assumptions and notation introduced thus far, we have

$$\{\int_E [|K_{V,C}(\bar{c}_b+\bar{\xi}_b;\bar{c}_a+\bar{\xi}_a)| - |K_{V,0}(\bar{\xi}_b;\bar{\xi}_a)|]^2 d\bar{\xi}_b\}^{1/2} \leq \tag{21}$$

$$\exp\{\epsilon/\hbar \int_{t_a}^{t_b} dt \|O\bar{c}(t)\|_2\} - 1$$

where the operator $O = \mu(x,y,z)\partial^2/\partial t^2 - K\nabla^2$,

$$\|O\bar{c}(t)\|_2 = [\int d^3x \sum_{i=1}^{3} (\mu c_{i,00} - Kc_{i,jj})^2]^{1/2} \tag{22}$$

and the region of integration in the left hand side of Eqn. (21) is $E = \{\bar{\xi}: \|\bar{\xi}\|_2 \leq \epsilon\}$.

We omit the proof of this theorem since it is closely analogous to the proof of theorem 2. The theorem indicates that if the channel function c(t,x,y,z) nearly satisfies the wave equation (16), then the probability that the corresponding channel C will be selected is comparable with the probability that a straight channel representing an undisturbed medium will be selected.

In this and other similar models we can regard the bundle of paths following the channel as objectively real. The functions defining the channel (such as $\bar{c}(t)$) correspond to the macroscopic aspect of this reality, and the interfering paths in the bundle can be viewed as its purely quantum mechanical aspect. The overall path of the bundle is determined by trans-temporal selection, subject to the requirement that the quantum mechanical laws are closely followed.

Although this model allows us to contemplate an objective picture of physical reality, it also entails a distinction between macroscopic and quantum mechanical levels similar to that stressed by Bohr. Here the macroscopic aspect defined by the channel is not directly tied in with particular acts of measurement as it is in standard presentations of quantum mechanics. However, the values of the macroscopic channel parameters at any given time can be regarded as measurements.

8. RELATIVITY AND NONLOCAL EFFECTS

We have not yet worked out the trans-temporal approach to quantum mechanics for relativistic field theories, but it should be possible to do this in a straightforward fashion. To define a channel we could adopt the approach used with the elastic solid in Section 7 and use functions $c_i(t,x,y,z)$ to specify the macroscopic characteristics of the fields. Trans-temporal selection of the channel can be defined in terms of a field theoretic path integral restricted to paths lying within the channel. The action for such an integral will be relativistically covariant, and the Lorentz transformation of a channel is simply another channel. Thus the trans-temporal selection process will conform closely to the quantum mechanical laws defined by the particular action functional. As a result, the channel may develop correlations between regions that have a space-like separation. However, this will not violate the relativistic rule that no signal can propagate faster than the speed of light. This rule applies to processes involving cause and effect within space-time, and does not apply to the process of trans-temporal selection which operates globally on the entire space-time situation.

The EPR effect, originally described by Einstein, Podolsky, and Rosen,[19] is a famous example of how a quantum mechanical system can exhibit correlated effects spanning large distances. In the version devised by Bohm[20] two electrons with correlated but otherwise indefinite spins begin to separate at a space-time point A. Later the two electrons pass through space-time points B and C, respectively. When observers at these points measure their spins in a particular direction, they find these spins to be correlated, even though each electron had individually undefined spin. This leads to the rather unexpected conclusion that the two electrons must be treated as a unit, no matter how great their spatial separation may be. The standard quantum mechanical analysis posits a wave function for the two electrons which collapses as a unit, and converts the electrons' initial state of undefined spin into a state with definite, oppositely directed spins.

The same effect is also predicted by the trans-temporal model of quantum mechanics. In this model, however, the collapse of the wave function is replaced by the trans-temporal selection of the channel. The physical set up of the electrons, the observers, and their instruments for measuring spin is such that the chosen channel can bend in one of two ways: with the first electron registering spin up and the second registering spin down, or with the opposite situation. Of course, the physical set up itself is an aspect of the path followed by the channel. Thus, if the channel followed a path in which the instruments measured spin in perpendicular directions, then it would be free to bend in four different ways, representing the four possible independent outcomes of these measurements.

Thus nonlocal effects can be understood as constraints on the ways that the selected channel can bend. We note that the trans-temporal formulation of classical mechanics also entails similar effects. If the chosen trajectory is curved so that particles in one location move in a certain way, then there must be correlated movements at other locations. This is due to the fact that in this trajectory the laws of motion are very nearly followed, and thus movements at separated locations B and C may be mutually influenced by earlier events at A.

9. PARAPSYCHOLOGY

A number of physicists have developed models of parapsychological phenomena based on the collapse of the quantum mechanical wave function.[3-5,21] In these models it is generally proposed that states of conscious volition can influence wave function collapse. This idea can be used to explain how the conscious self can direct brain processes, and it can also be used to account for some parapsychological phenomena such as psychokinesis.

The physicist Helmut Schmidt has performed a number of experiments suggesting that the will of a human observer can influence the outcome of subatomic events such as radioactive decay.(22) In a typical experiment, a random number generator (REG) based on radioactive decay is used to control a display consisting of a number of lights in a ring. Only one light in the ring is lit at any one time, and the apparatus is arranged so that this light will seem to perform a step by step random walk around the circle, with a fifty-fifty chance of going clockwise or counter-clockwise at any one step. Schmidt found in some cases that if a person observing the light desired it to move around the circle in a particular direction, then its random movements would show a small but definite bias in that direction over a considerable period of time. (In other cases the light might show a perverse tendency to move consistently in the opposite direction, and in still others its motion would apparently be completely random.)

Other investigators have observed similar phenomena. For example, a group headed by Robert Jahn of the Princeton School of Engineering has carried out an extensive series of experiments involving REGs driven by microelectronic noise.(23,24) These experiments also showed a consistent correlation between the volition of human observers and deviations in the expected long term behavior of random processes.

A striking feature of these experiments is that the human observer doing the willing has no conscious awareness of how subatomic quantum phenomena are being amplified within the experimental apparatus and used to generate the visible display. Furthermore, these subatomic events may even occur before the observer participates in the experiment and decides to desire a particular pattern of events. Thus Schmidt has obtained positive results from experiments in which the output of the random number generator was recorded on tape and played back later to operate the display.[22] This suggests that the observer's role is simply to desire to see a certain phenomenon, and that physical processes involving quantum mechanical randomness are unfolding teleologically so as to conform with the observer's desire.[24,25]

The trans-temporal approach to quantum mechanics provides a natural explanation for this kind of teleological behavior. To graphically explain this, it is useful to imagine the channel as a kind of flexible rod that can be trans-temporally deformed, but which tends to take on its own preferred shape. Let us suppose that the trans-temporal selection of the channel can be influenced by the will of individual persons. Suppose further that this will acts directly to attain a certain immediate outcome at time t_0 by inducing the channel function $\bar{c}(t)$ to take on values in some appropriate set W at this time. This could be modeled by modifying the function F of Eqn. (9) in such a way as to favor the selection of channels for which $\bar{c}(t_0) \in W$. It follows that if the channel is selected so as to nearly satisfy the physical laws, then the channel function $\bar{c}(t)$ for $t<t_0$ must vary in such a way as to lead up to some point in W. This natural flexing of the channel generates teleological behavior.

If the selected channel represents deterministic, machine-like interactions over the time interval from t_1 to t_2, then that section of the channel will be free to take on a limited variety of shapes, and we can think of it as being inflexible. But if it represents nondeterministic interactions (such as radioactive decay) in the vicinity of t_1, then the channel will be flexible at that point. Thus if an attempt is made to trans-temporally force the channel to take on certain characteristics at time t_2, the result may be the selection of a channel that bends abnormally t_1 so as to allow for this. Thus, if an observer in an REG experiment wills to see certain phenomena at time t_2, it may turn out that radioactive decays violating quantum mechanical statistics occur teleologically at time t_1 in such a way that the deterministic machinery of the experiment can produce the desired result by time t_2.

Effects of this kind do not depend crucially on strictly quantum mechanical phenomena. Sections of the channel corresponding to phenomena of "deterministic chaos" should also be quite flexible, as we argued in Section 3. Indeed, Rhine's classical experiments with dice, and also Jahn's experiments with cascading balls(24) may provide examples of psychokinesis involving deterministic chaos.

Costa de Beauregard(4) has pointed out that if it is

possible for a person to psychically influence the EPR spin observations at observation point B, then it follows that the observations at C must be similarly influenced. This allows for communication that could exceed the speed of light. According to the approach taken here, the psychical influences at B can be modeled by constructing F to favor certain configurations at B. As a result, channels with correlated effects at B and C will be preferentially selected, and the person's volition will also effect the situation at C. Since this kind of communication depends solely on trans-temporal effects, it does not violate the theory of relativity, which bars only processes of super-luminal communication based on cause and effect in space-time. We also note that this kind of communication does not depend specifically on quantum mechanical effects. We would also expect it to occur if a classical "chaotic" process at A resulted in correlated effects at B and C, and a psychic was able to influence these effects at B.

10. CONCLUSION

In this paper we have presented a basic framework for theories of physics in which the space-time history of events is seen as an objectively real continuum, and the laws of physics are seen as rules which apply globally to this continuum, constraining its form but not determining that form rigidly. We have applied this framework to both classical mechanics and quantum mechanics, and thus shown how classical mechanics can be endowed with an element of indeterminism, and quantum mechanics can be viewed as describing an observable objective reality.

In the case of quantum mechanics the postulated real continuum consists of a bundle of space-time paths constrained to lie within an envelope that we refer to as a "channel". The channel defines the macroscopic aspects of physical reality, and the paths (or alternatively, waves) within the channel define its microscopic or strictly quantum mechanical aspects. Physically permissible channels are specified by means of probabilities calculated using a Feynman path integral. By this means the laws of physics, expressed by the action functional of the path integral, are used to give a global definition of the channels that are allowed in nature.

In the case of classical mechanics the situation is much simpler. The postulated real continuum consists of a single classical path, and physically permissible paths are defined by the criterion that sensitivity of the action functional to variations in the path should be very small. We show that this definition allows for an enormous variety of paths, even though this sensitivity is made so small that no observable deviations from the classical equations of motion are allowed. Thus classical mechanics can be interpreted as a highly non-deterministc theory, and it also can be concluded that quantum mechanics can be highly non-deterministic when applied to apparently classical systems.

Traditionally, statistical laws have been incorporated in physical theories through the introduction of probabilistic postulates that are added to the laws of motion. Here we also use this approach, and we show that one basic probabilistic postulate can be used for both classical and quantum mechanics. In both cases the laws of motion (defined by the action functional) are used to provide probabilities, and the actual history of events is selected by a random choice in accordance with these probabilities. Since this choice is made once for all of space-time, we refer to it as trans-temporal selection.

To obtain the statistical laws of physics, it is not enough to define possible histories using the action functional, and then choose one at random. For example, this will not provide for the "arrow of time" defining the time direction in which the overall entropy of a system tends to increase. We note that this can be provided for, however, by systematically modifying the probabilities determined by the laws. We do this by multiplying these probabilities by a function F that favors certain possible paths over others.

If such an F is suitably defined, it can cause the selected space-time history to exhibit a variety of effects. For example, it may start out in an ordered state and show progressive increase of disorder; it may show an overall trend from a disordered state to an ordered state; or it may show an overall increase in disorder accompanied by the apparently spontaneous emergence of order at various times. In the latter case there may be teleological behavior, in which apparently random events conspire to produce a later organized event, and there may be non-local effects in which correlated events violating standard statistical laws occur in widely separated regions. These effects can occur in both the classical and the quantum mechanical models, and they depend on the effect of F in modifying the trans-temporal selection process.

Some effects of this kind have been observed in parapsychological experiments in which observers tried to mentally influence quantum mechanical phenomena. According to the trans-temporal selection model, it is possible to account for these experimental results by supposing that individual volition plays a role in the selection process. According to this idea, F should be constructed in such a way as to favor various sets of circumstances that are willed by individuals. The selection process chooses an overall history that blends together the various desired effects and also adheres as closely as possible to the laws of physics.

We should mention that these considerations are consistent with the idea that the trans-temporal selection process is carried out by a transcendental conscious agency that generates the physical world in accordance with basic non-deterministic laws, and the desires of individual, localized conscious entities. This idea has many ramifications, and it has particular relevance to the mind-body problem. It also provides an alternative to the philosophy of deism, which holds that God sustains the universe but plays no active role in the course of material events.

The trans-temporal model of classical and quantum mechanics is intended as a tentative suggestion that may hopefully lead to further insights into the nature of physical laws. In order to fully develop this model in its present form, much mathematical work is needed. For example, the model should be worked out in detail for relativistic field theories, and the proper way to define the channel for general systems should be investigated. It would also be particularly interesting to work out in greater detail the predictions this model makes for various teleological effects. These predictions might provide the basis for further parapsychological experiments which would test the trans-temporal model.

APPENDIX 1. Proof of theorem 1.

Here we begin with the arrangement described in the theorem. Place $\bar{x}_0$ midway between $\bar{z}_1$ and $\bar{z}_2$. The path of the moving disk corresponds to the path of a point which moves on straight lines in the area between the disks of radius 2r centered on the lattice points in the plane, and which reflects from these disks according to the rule that the angle of incidence equals the angle of reflection. Let $[\theta_a, \theta_b]$ be an interval of initial angles for the motion of this point, so that the rays emanating from $\bar{x}_0$ at these angles all strike the disk of radius 2r centered at $\bar{z}_1$, and the rays with angles θ_a and θ_b are tangent to this circle at either side.

We assume that in a grazing collision, the direction of the moving disk is unchanged. After the reflection from the disk at $\bar{z}_1$, the ray will pass through the region outside of the "shadow" S_1 marked in Fig. 4. By picking a suitable $\theta \in [\theta_a, \theta_b]$, this ray can be made to pass through any point in this region. Thus we can narrow down $[\theta_a, \theta_b]$ to an interval $[\theta_a', \theta_b']$ so that if θ lies in this interval, then the ray reflected from disk $\bar{z}_1$ meets disk $\bar{z}_2$, with grazing collisions for $\theta = \theta_a'$ and θ_b'.

After reflecting from disk $\bar{z}_2$, the rays with these initial angles will pass through all points not lying in the shadow S_2 marked in Fig. 4. Since the disk $\bar{z}_3$ does not intersect this shadow, we can narrow down $[\theta_a', \theta_b']$ to an interval $[\theta_a'', \theta_b'']$ so that for θ in this interval, the rays reflected from disk $\bar{z}_2$ will strike disk $\bar{z}_3$, with grazing collisions at $\theta = \theta_a$" and θ_b". The rays reflecting from disk $\bar{z}_3$ will then pass through all points not in the shadow S_3.

In general, we can find a sequence of nested intervals $[\theta_a(k),\theta_b(k)]$ such that the rays with initial θ in this interval strike disks $\bar{z}_1,\ldots,\bar{z}_k$ and precisely bracket disk $\bar{z}_{k+1}$ after reflecting from disk $\bar{z}_k$. This can be done if the radius 2r of the disks is small enough so that the shadow S_k cast by disk $\bar{z}_k$ never intersects any of the 8 disks immediately surrounding $\bar{z}_k$ other than the one opposite disk $\bar{z}_{k-1}$.

By choosing an initial angle of $\theta_0 \in [\theta_a(n-1),\theta_b(n-1)]$ we obtain the path required for the moving disk. Q.E.D.

APPENDIX 2. Proof of theorem 2. Some algebraic manipulations show that

$$K_{V,C}(\bar{z}_b;\bar{z}_a) = \exp[(i/\hbar)(A(\bar{c})+\dot{\bar{c}}_b\cdot\bar{\xi}_b-\dot{\bar{c}}_a\cdot\bar{\xi}_a)]K_{U,0}(\bar{\xi}_b;\bar{\xi}_a) \tag{23}$$

where $\bar{\xi}_a=\bar{z}_a-\bar{c}(t_a)$, $\bar{\xi}_b=\bar{z}_b-\bar{c}(t_b)$, $\dot{\bar{c}}_b=\dot{\bar{c}}(t_b)$, $\dot{\bar{c}}_a=\dot{\bar{c}}(t_a)$, and the potential U is defined by

$$U(\bar{\xi},t) = m\ddot{\bar{c}}(t)\cdot\bar{\xi} + V(\bar{c}(t)+\bar{\xi},t) - V(\bar{c}(t),t) \tag{24}$$

We can express $K_{U,0}(\bar{\xi}_b;\bar{\xi}_a)$ as

$$K_{U,0}(\bar{\xi}_b;\bar{\xi}_a) = (Q\varphi)(\bar{\xi}_b) \tag{25}$$

where φ is the wave function concentrated at the point $\bar{\xi}_a$ (i.e. $\varphi = \delta_{\bar{\xi}_a}$), and Q is the operator for propagation of a wave function from t_a to t_b through a straight channel ($\bar{c} \equiv 0$) using the potential U. We can also consider

$$K_{0,0}(\bar{\xi}_b;\bar{\xi}_a) = (Q_0\Psi)(\bar{\xi}_b) \tag{26}$$

where Q_0 is the corresponding operator for propagation through a straight channel with 0 potential.

Thus the left hand side of Eqn. (15) is

$$\left[\int_{[-a,a]^n} \left|\,|K_{U,0}(\bar{\xi}_b;\bar{\xi}_a)| - |K_{0,0}(\bar{\xi}_b;\bar{\xi}_a)|\,\right|^2 d\bar{\xi}_b\right]^{1/2} \leq \|Q\Psi - Q_0\Psi\| \tag{27}$$

The operators Q and Q_0 can be approximated as follows. Pick a $j>1$, define $\epsilon = (t_b-t_a)/(j-1)$, and let $t_k=t_a+(k-1)\epsilon$ for $k=1,\ldots,j$. Define the following two operators on functions of $\bar{\xi}$:

$$W = \exp[-(i\hbar\epsilon/2m)\nabla^2] \tag{28}$$

$$D_k(\bar{\xi}) = \exp[-(i\epsilon/\hbar)U(\bar{\xi},t_k)] - 1 \tag{29}$$

where ∇^2 is the n-dimensional Laplacian with respect to $\bar{\xi}$. Also let the projection operator P be defined by $P(\bar{\xi})=1$ if $|\xi_i|\leq a$ for $i=1,\ldots,n$ and $P(\bar{\xi})=0$ otherwise.

Using these operators we can approximate Q by

$$Q(j) = (I+D_j)PW\ldots(I+D_1)PW \tag{30}$$

We can also approximate Q_0 by $Q_0(j) = (PW)^j$.

Let ϕ be an arbitrary wave function. Then $\|U\phi\| = \|\phi\|$ since U is unitary. Also $\|P\phi\| \leq \|\phi\|$ since P is a projection. Finally, for $\phi_1 = P\phi$, $\|D_k\phi_1\| \leq \delta_k\|\phi_1\|$ where δ_k is the sup of $D_k(y)$ for $-a \leq y_i \leq a$, $i=1,\ldots,n$. By expanding Eqn. (30) and using these relationships, we can show that

$$\|Q(j)\Phi-Q_0(j)\Phi\| \leq [(1+\delta_j)\ldots(1+\delta_1)-1]\|\Phi\| \qquad (31)$$

for any wave function Φ. By taking limits as $j\text{---}>\infty$ we obtain

$$\|Q(j)\Psi-Q_0(j)\Psi\| \text{ ---> } \|Q\Psi-Q_0\Psi\| \qquad (32)$$

and

$$(1+\delta_j)\ldots(1+\delta_1) \text{ ---> } \exp[\int_{t_a}^{t_b} \sup_{\bar{y}}|U(\bar{y},t)|dt/\hbar] \qquad (33)$$

Combining these two results with Eqn's. (27) and (31) we obtain the conclusion of the theorem. Q.E.D.

REFERENCES

1. S.L. Jaki, *The Relevance of Physics*, (Univ. of Chicago Press, 1970)

2. E.P. Wigner, "Physics and the Explanation of Life," *Found. of Phys.* 1, 35 (1970)

3. H.P. Stapp, "Consciousness and Values in the Quantum Universe,' *Found. of Phys.* 15, 35 (1985)

4. O. Costa de Beauregard, "Time Symmetry and Interpretation of Quantum Mechanics," *Found. of Phys.* 6, 539 (1976)

5. R.D. Mattuck and E.H. Walker, "The Action of Consciousness on Matter: A Quantum Mechanical Theory of Psychokinesis," in *The Iceland Papers,* A. Puharich, ed. (Essentia Research Associates, 1979), pp. 111-159.

6. E.P. Wigner, "Epistemological Perspective on Quantum Theory," in *Contemporary Research in the Foundations and Philosophy of Quantum Theory,* C.A.Hooker, ed. (Riedel Pub. Co., 1973)

7. W. Heisenberg, "The Representation of Nature in Contemporary Physics," *Daedalus*, 87, 99 (1958)

8. M. Henon in *Proceedings of the Summer School "Chaotic*

Behavior of Deterministic Systems" (Les Houches, July 1981), G. Iooss, R.H.G. Helleman, R. Stora, eds. (North-Holland, 1983)

9. J.E. Fromm and F.H. Harlow, "Numerical Solution of the Problem of Vortex Sheet Development," *Phys. of Fluids* 6, 975 (1963)

10. J. von Neumann, *Mathematical Foundations of Quantum Mechanics,* (Princeton Univ. Press, 1955), pp. 417-445.

11. A. Daneri, A. Loinger, and G.M. Prosperi, "Further Remarks on the Relations Between Statistical Mechanics and Quantum Theory of Measurement," *Nuovo Cim.* XLIV, 120 (1966)

12. J.M. Jauch, E.P. Wigner, and M.M. Yanase, "Some Comments Concerning Measurement in Quantum Mechanics," *Nuovo Cim.* XLVIII, 144 (1967)

13. D. Bohm and B.J. Hiley, "Measurement Understood Through the Quantum Potential Approach," *Found. of Phys.* 14, 255 (1984)

14. E. Nelson, *Quantum Fluctuations,* (Princeton Univ. Press, 1985)

15. B.S. DeWitt and N. Graham, *The Many-Worlds Interpretation of Quantum Mechanics,* (Princeton Univ. Press, 1973)

16. R.P. Feynman and A.R. Hibbs, *Quantum Mechanics and Path Integrals*, (McGraw Hill, 1965)

17. R.P. Feynman, "Space-Time Approach to Non-Relativistic Quantum Mechanics," *Rev. of Mod. Phys.* 20, 367 (1948)

18. C.M. DeWitt, "Feynman's Path Integral", *Commun. math. Phys.* 28, 47 (1972)

19. A. Einstein, B. Podolsky, and N. Rosen, "Can Quantum-Mechanical Description of Reality Be Considered Complete?", *Phys. Rev.* 47, 777, (1935)

20. D. Bohm, *Quantum Theory*, (Prentice Hall, 1951)

21. H. Schmidt, "Collapse of the State Vector and Psychokinetic Effect," *Found. of Phys.* 12, 565 (1982)

22. H. Schmidt, "Evidence for Direct Interaction between the Human Mind and External Quantum Processes," in *Proceedings of the International Conference on Cybernetics and Society*, (IEEE, Inc., 1977)

23. R.G. Jahn and B.J. Dunne, *An REG Experiment with Large Data Base Capability, III: Operator Related Anomalies*, (School of Engineering/Applied Science, Princeton Univ., 1984)

24. B.J. Dunne, R.G. Jahn, and R.D. Nelson, *Princeton Engineering Anomalies Research*, (School of Engineering/Applied Science, Princeton Univ., 1985)

25. R.G. Stanford, "Experimental Psychokinesis: A review from Diverse Perspectives," in *Handbook of Parapsychology*, B. Wolman, ed. (Van Nostrand Reinhold, 1977)

1. Deterministic chaos generated by the Henon potential for a total energy of E=1/6. [This figure is from ref. (8), p 90.]

2. A closed orbit generated by a particle moving in accordance with the Henon potential.

3. An example of a simple quantum mechanical channel, showing one of the system paths which it encloses.

4. Diagram showing the limits of amplification of variations in initial direction for the classical pinball system. See Appendix 1 for details.

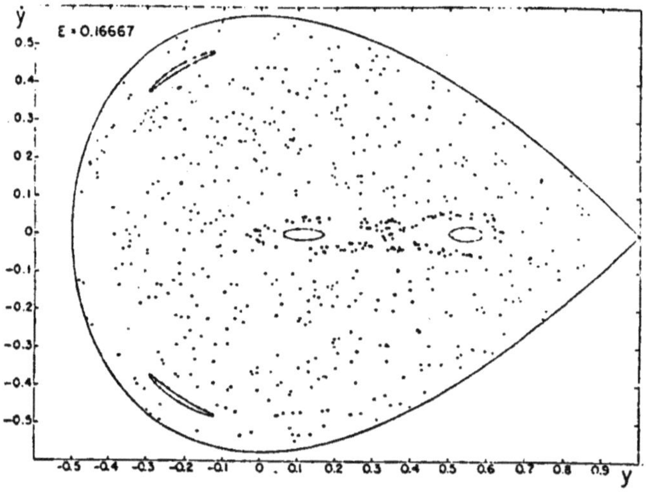

FIG. 1.

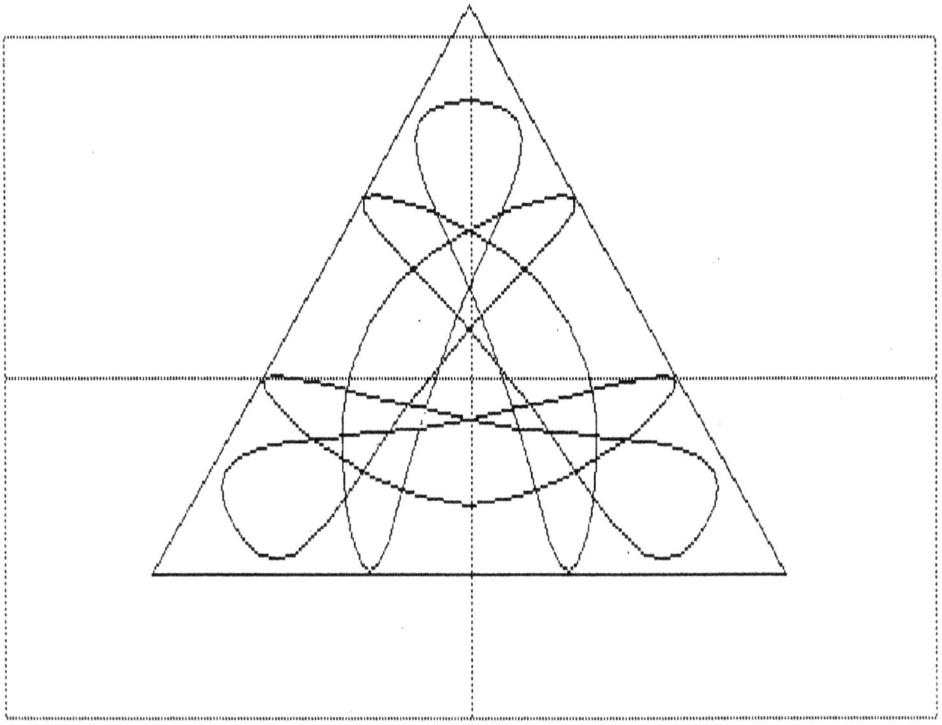

FIG. 2.

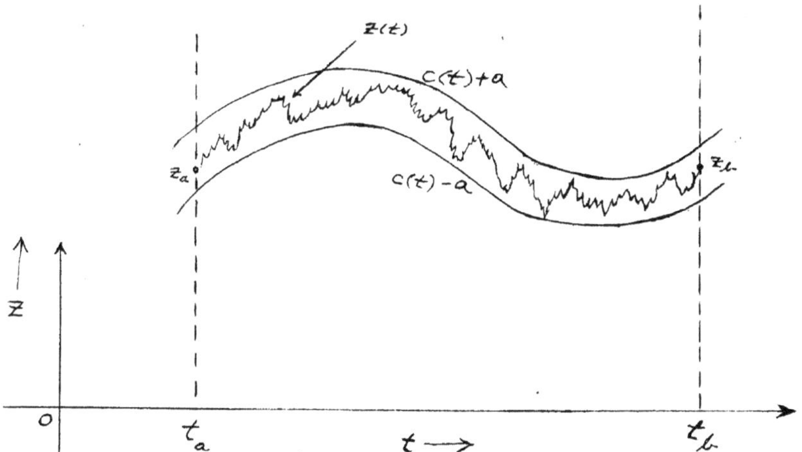

Fig. 3.

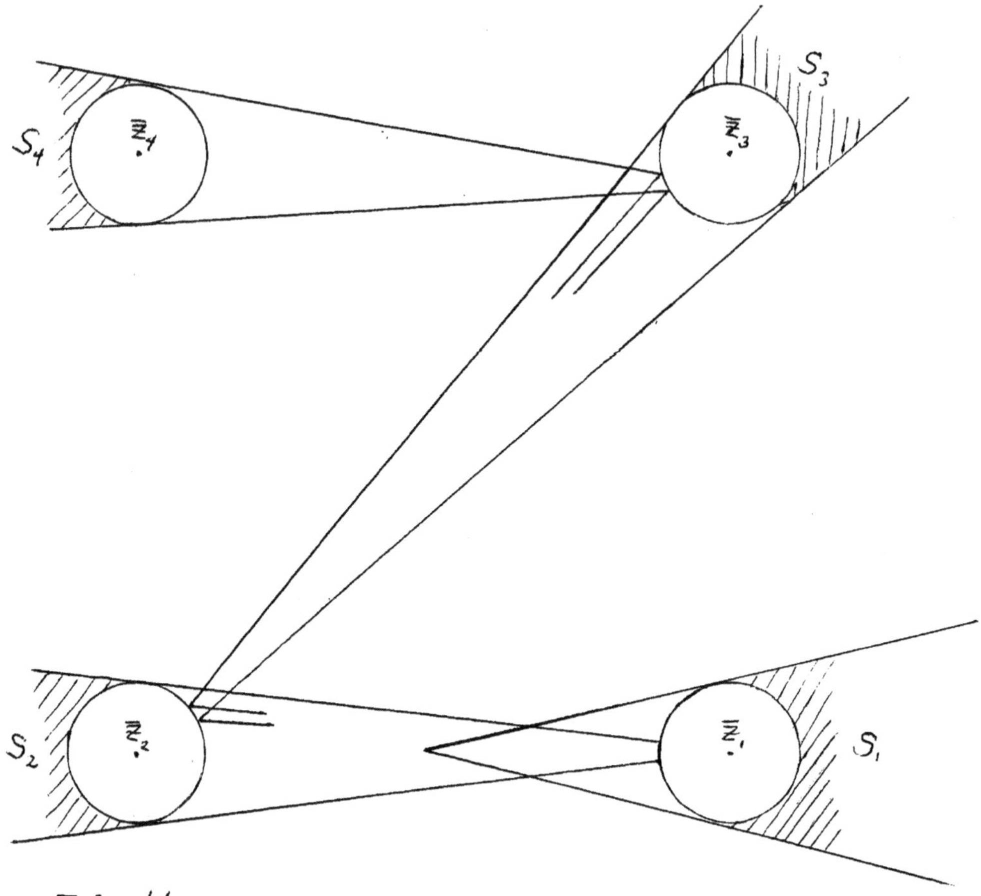

Fig. 4.

UNIVERSITY OF CAMBRIDGE

Department of Physics

From Professor B.D. Josephson, F.R.S.

Telephone: 0223-66477
Telex: 81292 CAVLAB G

Postal address:

Cavendish Laboratory,
Madingley Road,
CAMBRIDGE CB3 OHE,
England.

Feb. 12th., 1986.

Dr. Richard L. Thompson,
1136 Grand Ave., Pacific Beach,
San Diego, CA 92109,
U.S.A.

Dear Richard,

I'm sorry about the delay in sending my comments on your paper. Anyway, here's the list.

1) As I mentioned earlier, I think the comment on God creating the physical laws, etc. would be best not to be thrown at people at the beginning, though I don't see why it can't be put in the conclusion section at the end.

2) Feynman is misspelt on pp.4 and 39

3) On p.8 line 5 the word 'of' is omitted.

4) I don't think the detailed discussion of Henon's argument on p.9 et seq. is needed - you could just state the conclusions (I'm doubtful if it should be in even as Appendix I).

5) I'm not too sure that the argument based on time-reversal symmetry about systems being able to amplify small changes is valid as we may not be able to generate the time-reversed situation. Anyway, this seems to be another case where the result needed can just be stated. (It looks as if you have tried to put everything you have discovered on this subject somewhere in the paper, which leads to a comprehensive paper but not in general to a good one).

6) Top of p.18: the probability of a path is zero, so you must mean probability density.

7) It may be advisable to use a symbol other than 1 to denote the truth predicate (e.g. T).

8) The discussion on p.18 seems to make the choice of endpoints important, but this is unphysical. The time that is relevant presumably is the time-point at which selection operates in comparison with other times being discussed.

9) I see here (i.e. p.19) and elsewhere what seems to be to be some over-concern for what can be done rigorously, as if you are constantly concerned that someone will jump on you if you say something which you haven't rigorously proved. This very visible concern and attitude may actually detract from getting your message over.

10) I think that the detailed discussion of all approaches to the measurement problem is unnecessary and could be replaced by something much briefer.

11) On p.21, is "giving up the idea of objective reality" what you really mean here? I think that what people actually say has to be given up is the possibility of investigating reality itself as opposed to making measurements. (On points of principle I think you do really have to worry about people jumping on you).

12) p.22: if you are going to reference von Neumann's impossibility proof, you shouldn't do so without pointing out that it is fallacious.

13) On p.25 'the the' has been put where it should be 'to the'

14) p.26, l.8, to save confusion you might say '... where c in eqn. 10 is now confined to the set ... . Also I think that 'chosen in accordance with these probabliities' needs restating in a clearer form.

15) More confusing notation on p.27, with c being used for a new purpose in (12).

16) pp.35,36: maybe the account of PK should be condensed a little.

The paper is I think quite publishable in terms of its findings, though I'm not sure if the findings come over clearly enough as it is at the moment.

I was sorry to hear about your health problems in the letter I just got from you, and hope you are recovering now. Anyway, whenever you feel ready to give the lecture course let me know and I'll try to fix it up. I gather from some people I know who were there that the conference in Bombay was successful. Dipankar Home wanted to attend but was unable to get flights to Bombay.

Best wishes,

11, 12

Brian Josephson

CHAPTER 11

Interpretation and the *Śrīmad-Bhāgavatam*

Presented at the
Second Annual Conference of the
ISKCON Academy of Arts and Sciences
New Vrindaban, West Virginia / December 15–17, 2007

In this presentation I am going to talk about scientific interpretation of scriptures.

So to begin, let us consider the story of Lord Caitanya and the *ātmārāma* verse, for which he gave some sixty-one interpretations. At the same time, he was correcting interpretations of Sārvabhauma Bhaṭṭācārya that he did not approve of (Figure 1).

Figure 1: "The Liberation of Sārvabhauma Bhaṭṭācārya," *Śrī Caitanya-caritāmṛta, Madhya-līlā*, Chapter 6, Verses 187–190.

"After hearing the *ātmārāma* verse [SB 1.7.10], Sārvabhauma Bhaṭṭācārya addressed Śrī Caitanya Mahāprabhu, 'My dear Sir, please explain this verse. I have a great desire to hear Your explanation of it.'

"The Lord replied, 'First let Me hear your explanation. After that, I shall try to explain what little I know.'

"Sārvabhauma Bhaṭṭācārya then began to explain the *ātmārāma* verse, and according to the principles of logic, he put forward various propositions. The Bhaṭṭācārya explained the *ātmārāma* verse in nine different ways on the basis of scripture. After hearing his explanation, Śrī Caitanya Mahāprabhu, smiling a little, began to speak."

This suggests that on the one hand the *Bhāgavatam* has many valid natural interpretations, while at the same time, incorrect interpretations also. These interpretations are based on the background of knowledge behind the texts.

Turning to the subject of science, we find that the *śāstras* contain many statements that have a scientific impact. For example, one reason objections sometimes are raised to the theory of evolution is that *śāstra* does make statements about origins which have an impact on that period, suggesting the possibility that this period must be in need of some correction. At the same time, we often find that scientists and the scholars tend to reject scientific statements in the *śāstras* as primitive or just plain wrong. This is the general viewpoint.

What I would suggest is that there are many natural interpretations of the *śāstras* that make sense in the light of modern science. These interpretations can convey different scientifically clever ideas potentially of interest in presenting Kṛṣṇa consciousness to the academic and scientific world.

One pattern which I have seen empirically while looking at different examples is that these valid interpretations refer to subsets of information from the texts. In the given text, there may be two interpretations based on two different subsets of information from the texts, suggesting that a valid interpretation depends on reasonable meanings applied to a significant subset of the texts. I will illustrate that with some examples.

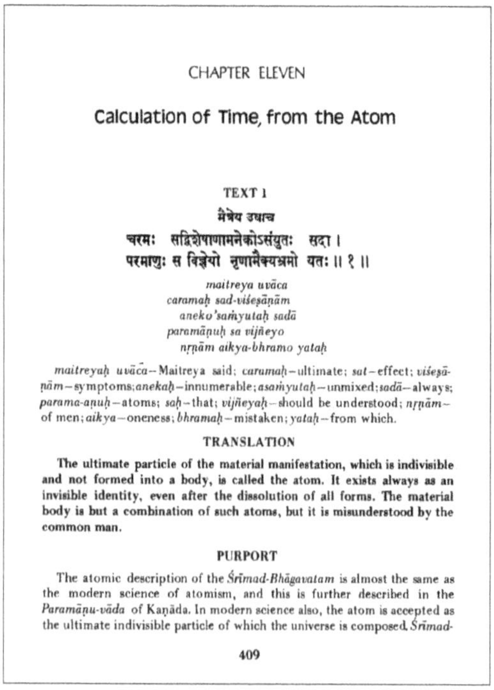
CHAPTER ELEVEN

Calculation of Time, from the Atom

TEXT 1

मैत्रेय उवाच
चरमः सद्विशेषाणामनेकोऽसंयुतः सदा ।
परमाणुः स विज्ञेयो नृणामैक्यभ्रमो यतः ॥ १ ॥

maitreya uvāca
caramaḥ sad-viśeṣāṇām
aneko 'saṁyutaḥ sadā
paramāṇuḥ sa vijñeyo
nṛṇām aikya-bhramo yataḥ

maitreyaḥ uvāca—Maitreya said; *caramaḥ*—ultimate; *sat*—effect; *viśeṣāṇām*—symptoms; *anekaḥ*—innumerable; *asaṁyutaḥ*—unmixed; *sadā*—always; *parama-aṇuḥ*—atoms; *saḥ*—that; *vijñeyaḥ*—should be understood; *nṛṇām*—of men; *aikya*—oneness; *bhramaḥ*—mistaken; *yataḥ*—from which.

TRANSLATION

The ultimate particle of the material manifestation, which is indivisible and not formed into a body, is called the atom. It exists always as an invisible identity, even after the dissolution of all forms. The material body is but a combination of such atoms, but it is misunderstood by the common man.

PURPORT

The atomic description of the *Śrīmad-Bhāgavatam* is almost the same as the modern science of atomism, and this is further described in the *Paramāṇu-vāda* of Kaṇāda. In modern science also, the atom is accepted as the ultimate indivisible particle of which the universe is composed. *Śrīmad-*

409

Figure 2: *Śrīmad-Bhāgavatam,* Canto Three, Chapter 11 "Calculation of Time, from the Atom."

The first example is taken from a chapter in *Śrīmad-Bhāgavatam,* Canto Three [Chapter 11] titled, "Calculation of Time, from the Atom" (Figure 2). There we find that a *truṭi* is defined as the time for integration of eighteen atoms into a molecule, and a numerical value for the *truṭi* is given. There we have $8/13{,}500$ seconds.

This is a very interesting statement because, on the one hand, it anticipates the idea of using an atomic clock to measure time. Present units of time are defined by atomic phenomena, and then are multiplied up to the larger units that we use, such as seconds and minutes and hours and so forth. Here we find the same idea is there in the *Bhāgavatam*. So this is an interesting scientific idea found in the text of the *śāstra*.

However, the length of the *truṭi* is a bit too long. Basically this *truṭi* comes to $1/1600$ roughly, and that will correspond to 1600 kilohertz. Now we all have our laptops here, most of which are working in the gigahertz range, or a billion computational cycles per second.

But it does not really make sense, [when comparing to the modern method], that the combining of eighteen atoms could take so long – $1/1600$ of a second. In fact, I get as a value for the modern *truṭi* the rather small number of $1/10^{12}$ [of a second]. In a modern science calculation that would be the approximate size of the *truṭi*.

So how can we consider this?

It seems that on one hand, the *śāstras* have given us a very clever scientific idea. On the other hand, the value of the *truṭi* is different. I would suggest, do not worry about the apparently oversized *truṭi*, and rather give credit to the text for offering an interesting scientific statement involving atomic clocks.

I would also suggest that you will find some other areas where this śāstric *truṭi* comes into play. One story involves the stealing of the calves and the cowherd boys by Brahmā. There we learn that one *truṭi* of Brahmā is equal to one earth year, and the basic equation is given.

Then there is another story in which King Kakudmī visits Brahmaloka, and then returns [to the earthly realm]. There it is said that he was in Brahmaloka for 27 times 4,320,000 *truṭis*. That would be years in the sense that one earth year is one *truṭi*. It would be that many *truṭis* [in this case roughly equivalent to 116.5 million earth years].

And then we find another interesting *truṭi* mentioned in the *Sūrya-siddhānta*. It is $1/33{,}750$ [of a second], which is different from the *Bhāgavatam* value.

If you put these three apparently disparate elements together you get an interesting result. Basically, if you multiply the number, 27 times 4,320,000, against the length of the *truṭi*, that would give you the number of seconds on Brahmaloka [for the time] that the King was visiting there. And it comes out to be 3,456 seconds [~58 minutes] when you multiply that out.

This is kind of interesting, because first of all, the King was there for about an hour [as experienced on Brahmaloka]. That makes sense in terms of the story, because it states that he listened to a musical performance, and then had a brief conversation with Lord Brahmā. An hour would be 3,600 seconds, and now we have this number of seconds which is very close to it.

The other interesting thing is that the number 3-4-5-6 looks rather ordered [*laughter*] ... It does not look like a product of chance.

I would suggest that something is going on here. It could suggest people were making calculations involving units of time on Brahmaloka, and this is the value that they came up with. In other words, this is a product of design, not of random chance.

Another interesting thing to note, is that if we do this backwards and consider that the King was on Brahmaloka for an hour, and then ask how long would a *truṭi* be, we also get something close to this value. And this *truṭi* is 1/20 of the *Bhāgavatam truṭi*, suggesting a relationship to the *Bhāgavatam truṭi* with a ratio of 1 to 20.

It thus seems that we have significant use for a *truṭi* that is about this size, about the size of the one in the *Bhāgavatam*. We find that the two figures are connected, namely with time on Brahmaloka. So that is basically a summary.

I would now like to turn to a different example involving cosmography described in the Fifth Canto of the *Bhāgavatam*. I am going to start with an interpretation given by a South Indian gentleman back in the 1970s. Basically he was wondering: If we are on an earth globe about 8000 miles in diameter, and the *Bhāgavatam* is describing these huge landmasses and gigantic mountains in Bhū-maṇḍala, how then would we relate these two things? He had the idea that the earth was sort of mounted on a mountain, sticking up from the base of Bhū-maṇḍala. This was his proposal. And the same idea has been recently revived.

The first thing I will say about this, is that this is an interpretation. I would tend to think that this is perhaps not a very good interpretation, because it is not based on solid scriptural evidence.

It also leads to many practical problems, such as if you stand on something that is floating above these gigantic land masses, why then can't you see them? I suppose someone could say, "Well, they are invisible." So that becomes a challenge.

There is another interpretation of Bhū-maṇḍala that we know of historically, given in the *Siddhānta-śiromaṇi* written by Bhāskarācārya in about the eleventh century. Bhāskarācārya wrote that most learned astronomers have stated that Jambūdvīpa embraces the whole Northern Hemisphere lying north of the South Sea, and that the other six *dvīpas* and the seven seas [described in the *Bhāgavatam*] are all situated in the Southern Hemisphere. [Note: this period lacked regular navigation to the "South Sea," south of the equator.] It is here that he lists these seas and *dvīpas*.

Bhāskarācārya is saying that Jambūdvīpa is the Northern Hemisphere of the earth, and that most learned astronomers in India held this viewpoint. For example, here we have an image of a globe – this globe, by the way, was commissioned by Mahārāja Jai Singh of Jaipur [during the early 18th century] – where you see that the Northern Hemisphere basically has identifiable mountain ranges and so forth of Jambūdvīpa imprinted on it (Figure 3). This area here would be Bhārata-varṣa. And then down in the Southern Hemisphere, you can see alternating oceans and *dvīpas*. So we have historical evidence that this interpretation really was held.

Figure 3: Globe commissioned by Mahārāja Sawai Jai Singh II c. 1730s, showing Jambūdvīpa in the Northern Hemisphere and the other six *dvīpas* in the Southern.

Here is an enlargement of the image showing some of the observable features of Bhārata-varṣa. For example, here is Delhi, here is Prayāga, here is Kāshī, here is Gayā, that is Jagannātha Purī, and so forth. Also we have Chin and Mahachin, referring to China and greater China. Over here we have Mecca [Makkah] (Figure 4).

So this offers an interpretation. The interesting thing about this interpretation is that here the earth globe *is* Bhū-maṇḍala. It is not that it is floating next to Bhū-maṇḍala, but rather that Bhū-maṇḍala is taken to be a globe.

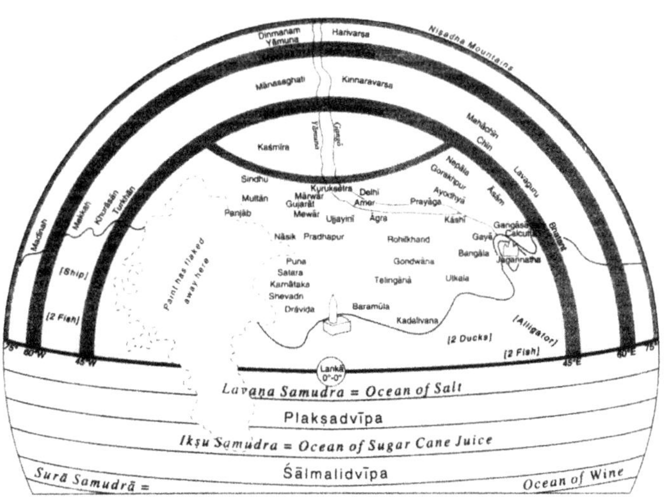

Figure 4: English transliteration of identifiable geographic locations associated with Jambūdvīpa described in Sanskrit on the globe commissioned by Mahārāja Sawai Jai Singh II, c. 1730s (Figure 3). See Thompson's *Mysteries of the Sacred Universe* (2000), p. 50 for more information. MSU in turn credits Joseph Schwartzberg's essay, "Introduction to South Asian Cartography," published in *Cartography in the Traditional Islamic and South Asian Societies* (1987) by the University of Chicago Press, pp. 396–97.

One might raise some objections to this. The first objection might be: If the *Bhāgavatam* describes Bhū-maṇḍala as a flat plate, then how could it be a globe?

There is a scholar named Randolph Kloetzli, who did a study of the cosmology of the *Viṣṇu Purāṇa*, and he went into this in great detail. He argued that the flat plate of Bhū-maṇḍala, as described in the *Viṣṇu Purāṇa*, corresponds to an astrolabe. An astrolabe is an instrument that was still being used during the time of Columbus (image 5).

Figure 5: Mercator's astrolabe (1545), in the Moravska Gallery, Brno, Czech Republic.

In an astrolabe, you take the earth globe, and you map it onto a flat plate. Here we have a picture showing how that works (Figure 6a). If the flat plate is like this, with the pattern of Jambūdvīpa on it, the mapping by what is called stereographic projection wraps that around the globe (Figure 6b). Or if you start with a globe, you can then unwrap it and get your map on the plate.

This is nothing too mysterious, it is just a map projection. And today we have many different map projections for mapping the round globe onto a flat surface.

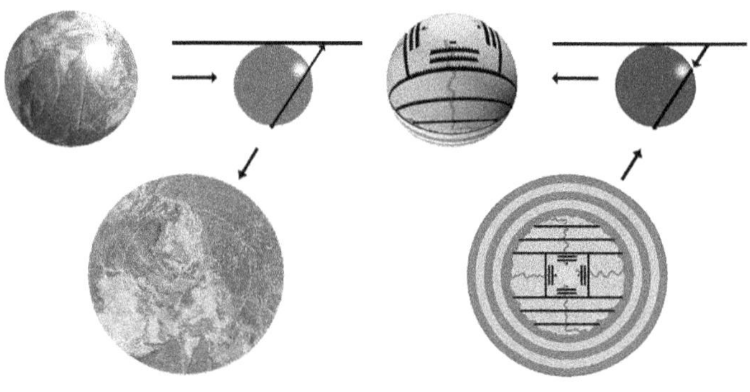

Figure 6a left: The earth globe projected stereographically onto a plane. **Figure 6b right:** The inverse-stereographic projection of Bhū-maṇḍala onto a globe.

In an astrolabe, this is in fact done. The globe is mapped onto a flat plate, and the flat plate is used for computations by moving other plates with little indicators on them. So, it is a kind of computational device (Figure 7). This could offer an explanation of the relationship between the earth globe and the Bhū-maṇḍala disk.

There is another point where one might have some misgivings about this particular globe model of Bhū-maṇḍala, and that is that the globe is not geographically realistic. For example, in the Northern Hemisphere, we do not see anything like Jambūdvīpa. Nor is there North America, Africa, and Europe, so on and so forth. We do not see any of this, [though there are features identified with Asia.]

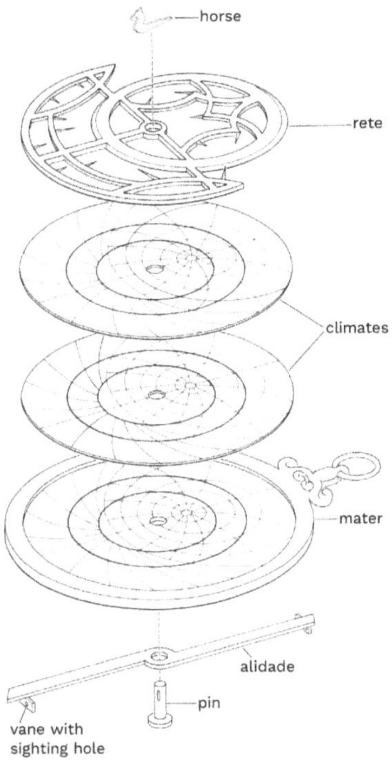

Figure 7: Exploded picture of an astrolabe, showing plates representing the earth (*mater*) and the ecliptic and star positions (*rete*). (MSU Fig. 3.15 credits Randolph Kloetzli, "Maps of Time–Mythologies of Descent: Scientific Instruments and the *Purāṇic* Cosmography," in *History of Religion* 25, No. 2, 1985.)

It is interesting that there is yet another interpretation of the text backed by quite a large amount of information, which correlates Jambūdvīpa with a region of Asia. It extends from Northern India up to regions of Siberia, as shown here. There is qualitative correspondence between the different mountain ranges we find on the actual globe and the different mountain chains of Jambūdvīpa shown in a somewhat more stylized fashion. In the *Vāyu Purāṇa* and the *Matsya Purāṇa* there is a lot of information backing up this particular interpretation (Figure 8).

So what we are finding are various interpretations of Bhū-maṇḍala and Jambūdvīpa which serve different purposes and are based on different sets of evidence taken from contextual tradition.

Now one point that we still have not addressed, and that is that the earth globe we are talking about is small in comparison to the Bhū-maṇḍala disk as described in the Fifth Canto. The earth globe is about 8,000 miles across – a figure based on the analysis of the *Siddhānta-śiromaṇi* and the *Sūrya-siddhānta* that agrees with modern evidence of the size of the globe. But if you look at Bhū-maṇḍala as described in the *Bhāgavatam*, you will find that it is four billion miles across. That is quite a discrepancy.

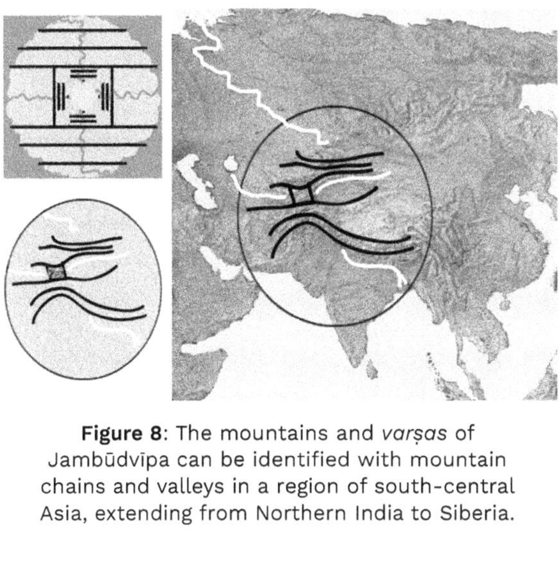

Figure 8: The mountains and *varṣas* of Jambūdvīpa can be identified with mountain chains and valleys in a region of south-central Asia, extending from Northern India to Siberia.

Here I want to turn to Bhū-maṇḍala as a large disk with the dimensions as described in the *Bhāgavatam*. This is an interesting picture relating it to the orbit of the planets. This picture was made by a certain South Indian swami who lived in the nineteenth century, Tiruveṅkaṭa Rāmānuja Jīyar Swami. While I was travelling through South India I met a scholar, Lakshmi Thathachar, who runs a Sanskrit institute [The Academy of Sanskrit Research] in Melkote, which is a center for Rāmānuja. When he learned of my interest in the Fifth Canto, he brought this diagram [offering a description of the Bhū-maṇḍala plane].

This is a small part of a much larger diagram. The purpose of the disk top – this part – is of interest. Here we have the earth globe, which interestingly enough is being shown next to Jambūdvīpa (Figure 9a).

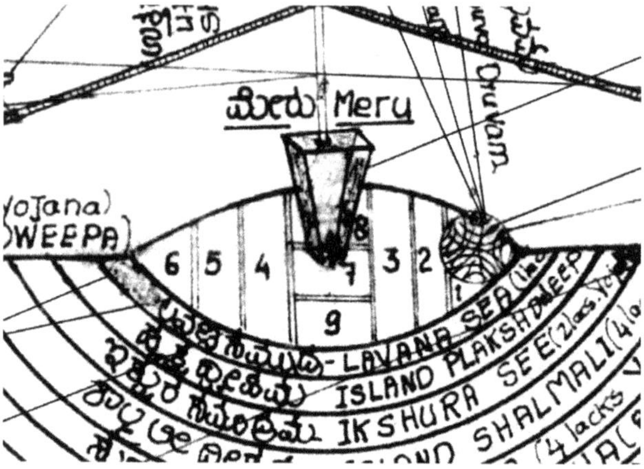

Figure 9a: In Tiruveṅkaṭa's diagram, the *varṣas* of Jambūdvīpa are numbered 1 through 9. Note that the earth globe is placed in Bhārata-varṣa (number 1).

Going around the earth, in a small orbit, is the moon. Actually, these planets can all be identified by the different figures shown above them. These go up to greater heights, as you can see from the lines. Here is the sun, here is the axis of the sun's chariot, here is Mount Meru in the center. This second object is the same sun at a different time, around to the other side of the figure. Here is the axis, here is the wheel of the chariot. Here you have the Mānasottara Mountain, which is like a circular racetrack the wheel is running on (Figure 9b).

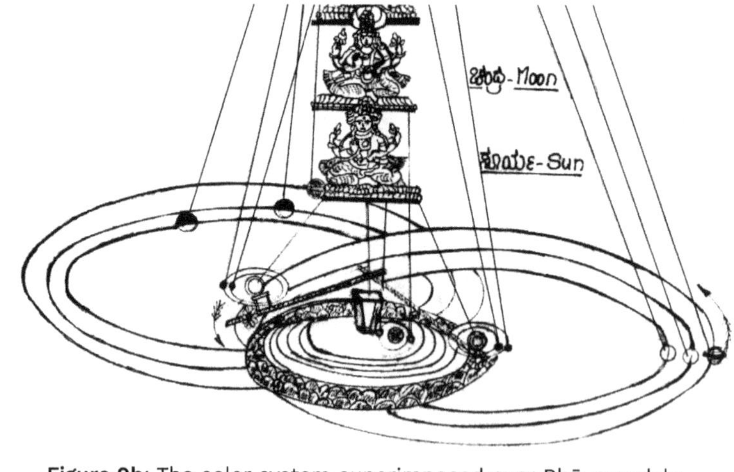

Figure 9b: The solar system superimposed over Bhū-maṇḍala. Note that the sun is shown orbiting the earth and moon, which are shown next to Mount Meru. Mercury and Venus are shown closely orbiting the sun, and Mars, Jupiter, and Saturn are orbiting further out. This is shown for two positions of the sun simultaneously.

The interesting thing in this diagram is that if this is the sun – you see the sun is going around the earth, which is in the center – but here are Mercury and Venus, and as you can see they are also orbiting the sun. Here they are again over here. As for Mars, Jupiter, and Saturn, they are also orbiting the sun. They are showing quite a bit larger orbits, and look at the little rings on Saturn that you can see!

So this is the diagram the Swami came up with back in the nineteenth century. Immediately you could see an interesting correlation between this diagram and a diagram made by a famous Danish astronomer, Tycho Brahe, during the 1500s. Tycho Brahe was both the most famous naked-eye astronomer that we know of, as well as the last major astronomer to consider the geocentric model of the solar system (Figure 9c).

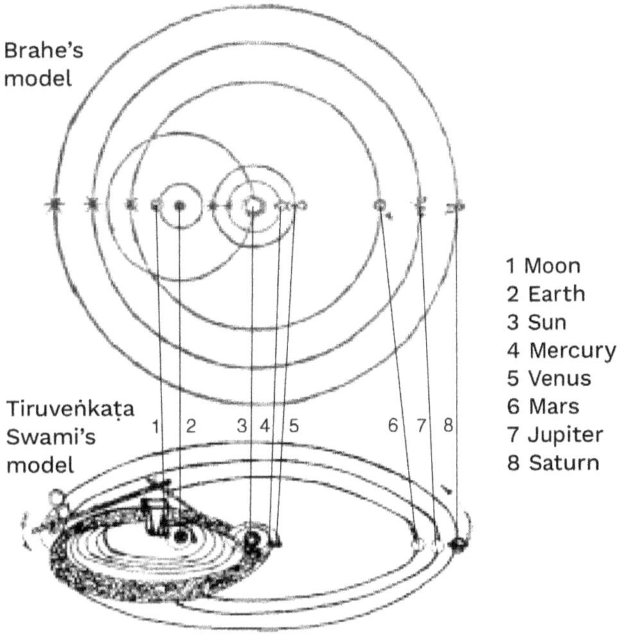

Figure 9c: Tiruveṅkaṭa Swami used Tycho Brahe's geocentric model (whether he knew it by that name or not) to interpret planetary motion in the *Purāṇas*.

In Brahe's model, here is the earth and the moon going around the earth. Here is the sun going around the earth, and here is Venus and Mercury going around the sun, with Mars, Jupiter, and Saturn also going around the sun. Everything correlates with the Swami's diagram, as you can see.

I would suggest that probably the Swami got his diagram from Tycho Brahe. After all, he lived in the nineteenth century, so it is plausible. But it is interesting to see what happens if you do what the Swami essentially did, where if you take Bhū-maṇḍala with dimensions given in the *Bhāgavatam*, and then superimpose that on the geocentric model of the solar system.

We can do better now than Tycho Brahe. Nowadays you can utilize an ephemeris software program, which is quite accurate. If you translate the position of the earth so that it remains fixed in the center, you will then get the orbits of the planets. You will see that the sun is going around the earth, with the various planets then going around the sun.

Moderator: We are getting toward the 5 minute limit.

Okay, I am going to skip a little bit ...

This is an illustration of what I just described using Mercury as an example – the blue spiral graph-like line is the geocentric orbit of Mercury (Figure 10). Basically, you have the sun going round the earth, and Mercury going round

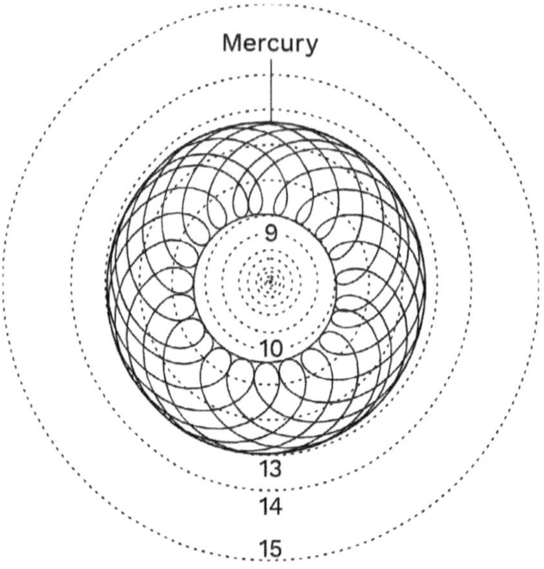

Figure 10: The geocentric orbit of Mercury, compared with Bhū-maṇḍala using 8.489 miles/*yojana*. The inner boundary of the orbit of Mercury swings in and nearly grazes feature 10 (inner edge of Śākadvīpa), and its outer boundary swings out and nearly grazes feature 13 (Mānasottara Mountain). We can sum this up by saying that Mercury's boundary curves are tangent to features 10 and 13.

the sun. Naturally you are going to get a spiral pattern like this, and that is what we see in blue here.

The interesting thing that you note with this pattern – and I guess I will pause here and give a little bit of the history behind this, because I found the Swami's diagram later. The way this idea originally came up for me was when Harikeśa wanted an arrangement showing the planets above Bhū-maṇḍala for the Temple of the Vedic Planetarium. He said, "Well, so show the planets over Bhū-maṇḍala," to which I responded, "I don't think that is going to work!"

But he kept repeating it on various occasions. So finally, one day I thought, "Okay I will do it and see what happens." I did not expect to get any results.

What I found when I did it was, as we can see here in the case of Mercury, that this annular curve of the spiral geocentric orbit is tangent to one of the ring structures of Bhū-maṇḍala, and the inner boundary curve is also a tangent to one of the ring structures of Bhū-maṇḍala.

Now the same is true of the other planets. There are quite a number of other details which all fit together to indicate that Bhū-maṇḍala could also be acting, in effect, as a map of the geocentric orbits of the planets. This is an interpretation that emerges with regard to the size of the features of Bhū-maṇḍala.

By the way, if you run your computer program long enough, the spirograph will fill in to give a solid doughnut shape, which is tangent to the outer curve at one point and to the inner curve at the other point. If we go from the small earth globe to the full-sized Bhū-maṇḍala, we find that those magnitudes do have a value, even though one has to leave them aside when considering just the small earth globe.

Another interesting correlation that emerges from this involves an examination of the *yojana* based on a historical study of Megasthenes, the Seleucid Greek ambassador to India [at the court of Candragupta of the Mauryan Empire], as well as some early Chinese Buddhist pilgrims. It indicates that there are basically two *yojanas*, one of which is about half the size of the other. It actually comes out to $^{16}/_{30}$ of the other *yojana* (Figure 11).

Well, if you go back to the orbit diagram, you will see that depending on how big a *yojana* is, this orbit, or equivalently the Bhū-maṇḍala map, will shrink or expand. The question then is: What is the best value for the *yojana*?

Using a root mean square test, what I found was that the best fit value for the *yojana* comes to 8.5 miles. Further historical study reveals information indicating that the ancient Egyptians also had a value for their units, which we can then relate to the *yojana*.

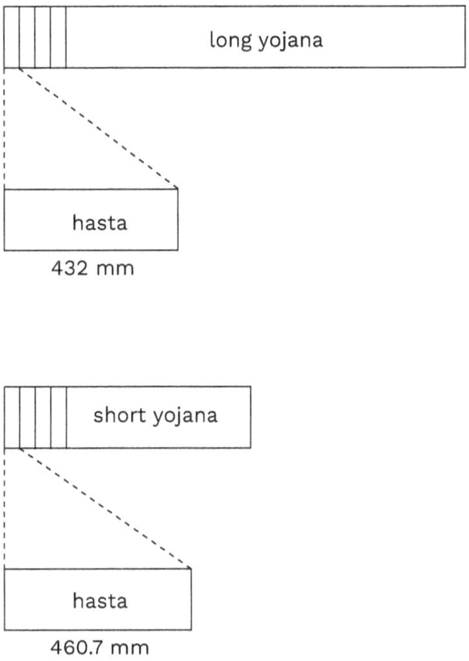

Figure 11: There are at least two *yojana* standards: A long *yojana* of 32,000 *hastas* and a short *yojana* of 16,000 *hastas*. The *hasta* of the short *yojana* is slightly longer than that of the long *yojana*.

Basically, they apparently were basing units on divisions of a degree of latitude. It turns out that an 8.5-mile *yojana* corresponds to 7.5 minutes of latitude as measured at the equator, while the other shorter *yojana* corresponds to exactly four minutes of latitude. One quarter of that is a unit called *krośa*, which is one unit of latitude, and that is the same as the nautical mile being used to this very day.

The point is that by taking a certain degree of flexibility in looking at different interpretations, and then basing these interpretations on good evidence, we can find a lot of interesting things in the *śāstra* that I think may be of interest to people in the academic world.

Of course, it is going to take a lot of work to be able to introduce this. Admittedly, I have a somewhat spotty publication experience in the academic world [with this material]. But I did see some interesting conniption fits [*laughter*] … as I presented some of these ideas [*laughter*] …

Now we are at zero minutes. I had to drop about three slides.

Moderator: Thank you very much [*applause*] … We can open it up for questions. Any questions or comments?

Question: I have a śāstric question. There appears to be credible information coming from these texts that is obviously old. What age would you give the *Bhāgavatam*? Even when considering the youngest age that you might want to give to it, we are dealing with information that was clearly not available in the scientific world until well after it was written. The question then is, how did these folks know?

Well, if they knew it by revelation [*unclear*] ...

Answer: There are those two basic possibilities that you mention.

One is if you postulate an ancient civilization older than the civilizations that we know of, in which people have this kind of scientific knowledge, and then somehow in a sort of imperfect way pieces of it came down to our own time.

Then the second is that a sage or yogi with connection with higher beings, acquired the knowledge. That would be an idea that you would find in India itself. For example, in the *Sūrya-siddhānta*, it is said that the text was communicated to Mayāsura by an emissary of the sun-god. The idea of getting it from higher sources is there in the Indian tradition.

I tend to work with the idea of an ancient civilization because of all these correlations with ancient metrology in Egypt, as well as correlations from around the world. This would seem to suggest that the people of the ancient past had access to a more advanced science, followed by a period of darkness and dark ages. Then to some extent people recovered from that in classical antiquity, followed by another set of dark ages that began to recover during the Renaissance. And that brings us to the present.

Question: Regarding the *truṭis* and the correspondence that can be made between the *Sūrya-siddhānta* version of the *truṭi* and the story of King Kakudmī, and the difference in time between our time and Brahmā's time. What can we get from that?

It seems that there is a coherence between the *Sūrya-siddhānta* version of the *truṭi* and the *Bhāgavata*. What I am not clear about is, did Prabhupāda discuss a way of understanding a correspondence between either the *Sūrya-siddhānta* or the *Bhāgavata's* definition of the *truṭi*, and the modern calculation as to how long that would take?

Also, you made this point about an even ratio of 1 to 20 between the *Sūrya-siddhānta* and the *Bhāgavatam*. I wonder what you make of that ratio?

Answer: Well, the basic way that I look at it, first of all, is that we are seeing a coherence between the cowherd boys' story, King Kakudmī's story, and the *truṭi* in the *Sūrya-siddhānta*. They all appear to mesh. And the fact that the number is 3456 suggests that this was done deliberately.

What does that have to do with the modern *truṭi*, $1/10^{12}$ of a second?

I would suggest that just as in the astronomical examples where we had a Bhū-maṇḍala model which ignored the very large distances, and then later we came to an interpretation which took into account these large distances, what I am arguing here is similar in the case of the *truṭi*. In the case of the *Bhāgavatam*, that is when to set aside the large *truṭi* that comes into play in connection with the *Sūrya-siddhānta* and the two stories.

Now there is one further point, namely the ratio of 1 to 20. Because the *truṭi* in the *Bhāgavatam* is not exactly the same as the *truṭi* in the *Sūrya-siddhānta* … it is off by a factor of 20.

If you directly use the *Bhāgavatam truṭi* in the King Kakudmī calculation [as compared to the *truṭi* in the *Sūrya-siddhānta*], then you would find that the king was there for about twenty hours. That would not be as good a fit from the point of view of the story. It would also obscure the 3456 number. So in that instance, the calculation would become much weaker.

Question: You have spent a lot of time trying to look at the *Bhāgavatam* in a scientific way and to show how it has scientific relevance. What is the point of doing that? Why do you do that?

Answer: Ultimately it has to do with facilitating people's faith, because for some reason the *Bhāgavatam* was written with reference to things that impinge on science. This was on a slide that I had on statements that would have an impact upon a scientific view of the world. Because those statements are there, you have to deal with them.

For example, I mentioned the point that the Vedic literature tends to go against the theory of evolution as generally understood by scientists. So here you are, trained up in believing evolution from an early age. I mean for me it began with pictures of the dinosaurs and so forth. I think I first got that when I was about seven. It's an early indoctrination [*laughter*] …

But just to continue with this example, you tend to come in contact with Kṛṣṇa consciousness with a thoroughgoing belief in evolution. I remember being on *harināma* parties in Manhattan. We were there with *mṛdaṅgas* and everything, and I found myself thinking: "What about the dinosaurs?" [*laughter*] …

So ultimately there needs to be an answer to these questions that satisfies people. There is the basic consideration that if the text is saying things that are outliers as far as science is concerned, then how can you put faith in the transcendental statements, which are beyond your present levels of realization?

Perhaps many of us do not have direct realization of these higher levels of *bhakti*. If we did, we could probably say, you know, to heck with science and all those different issues. I will simply sit under a tree like the Gosvāmīs, drink the nectar of Kṛṣṇa consciousness, and influence people through purity and effulgence and so forth.

But there can be challenges with faith. For example, I remember when the Fifth Canto first came out. I was in Atlanta at the time, and I was taking a walk through a nearby golf course, looking at the trees. I knew from the Fifth Canto there is a description of trees that are as large as the diameter of the earth. And I was looking at the trees on the golf course thinking, "Well, these trees are not that big. There is nothing I can do to force that. So how am I going to understand this?"

So ultimately you need to have some understanding. I think that would be the basic motivation, why science is important.

A synopsis of the presentation included in the
Conference Program for the Second Annual Conference
of the ISKCON Academy of Arts and Sciences
New Vrindaban, West Virginia | December 15–17, 2007

⋆ ⋆ ⋆

Interpretation and the *Śrīmad-Bhāgavatam*

Sadāpūta dāsa

Lord Caitanya is famous for giving sixty-one meanings of the *ātmārāma* verse to Sanātana Gosvāmī. Here I would like to offer some insights on meaning in the *Bhāgavatam* based on a scientific background.

Example 1. In the *Bhāgavatam*, the time for integration of 18 atoms is one *truṭi* or $^{8}/_{13,500}$ seconds. Conceptually, this agrees with the modern atomic definition of time, but the time interval is too long. The modern *truṭi* is about $^{1}/_{1000000000000}$ sec, approximately.

Instead of seeing a contradiction here, let us try accepting the basic idea, while allowing for different time intervals. This allows us to consider the *Sūrya-siddhānta truṭi* = $^{1}/_{33,750}$ seconds. Using $^{1}/_{33,750}$ seconds of Brahmā as a *truṭi* (moment) of Brahmā, and referring to the stories of King Kakudmī and the stealing of the cowherd boys, we can calculate that King Kakudmī spent 3,456 seconds of Brahmā in Brahmaloka, or about one hour (3600 seconds of Brahmā).

The time interval of 3,456 seconds fits the story of King Kakudmī, and it is surely no coincidence. So we see that something is going on behind the scenes here. Let us generalize on this: Statements in the *Bhāgavatam* may have a variety of natural meanings which may involve alterations in quantitative values.

Example 2. Picture by Bādarāyaṇa Murthy. Here we see the earth supported on a mountain extending up from Bhāratavarṣa in Bhū-maṇḍala. This is problematic, but *Siddhānta-śiromaṇi* (by Bhāskarācārya, 11th century) gives another way of relating the earth to Bhū-maṇḍala:

> "Most learned astronomers have stated that Jambūdvīpa embraces the whole Northern Hemisphere lying north of the salt sea; and that the other six *dvīpas* and the seven seas ... are all situated in the Southern Hemisphere."

This shows that "most learned astronomers" regarded the earth globe as Bhū-maṇḍala itself, not as an object situated next to Bhū-maṇḍala in space. But why would Bhū-maṇḍala be wrapped around the earth?

The scholar W. Randolf Kloetzli gave an answer in an article interpreting Bhū-maṇḍala as an astrolabe – a navigational computer based on a stereographic mapping of the earth's surface onto a flat plate. This shows that by allowing Bhāskarācārya's interpretation, and not rejecting it as contradictory, we are led to an interpretation of the Bhū-maṇḍala disk as a sophisticated computer.

But what can we say about the large magnitude of Bhū-maṇḍala in contrast with the small earth? Let us turn to an interpretation which takes this into account.

Diagram by Tiruveṅkaṭa Rāmānuja Jīyar Swami of the 19th century, that I discovered in the town of Melkote in India. Let's look at the planetary orbits shown in this diagram. Mercury, Venus, Mars, Jupiter, and Saturn are all shown orbiting the sun, and the sun is shown orbiting the earth. This, in fact, is the geocentric system of the famous Danish astronomer Tycho Brahe.

I have pointed out before that any planet can be taken as the center of motion in the solar system. In particular, the earth can be taken as the center of motion, and we can accept a geocentric model of the solar system. We have to be careful, however, to not claim that the earth is absolutely the center of motion after placing it in the center by a relativistic argument.

Consider the following relativistic statement of the heliocentric system: "Any planet can be taken as the fixed coordinate frame for all movement in the solar system. This planet is, by definition, stationary. The sun is seen to revolve around this planet, and the other planets are all seen to orbit around the sun." This statement is relativistic in the sense that it does not single out any planet for special attention. It includes the geocentric model of Tycho Brahe and Tiruveṅkaṭa Swami. It also includes, for example, a perspective in which the planet Venus is fixed and the earth orbits around the sun. Indeed, it contains five perspectives in which the earth orbits around the sun.

If we avoid making a relativistic argument and drawing an absolute conclusion, we must see Tycho's (and Tiruveṅkaṭa's) diagram as one instance of a relativistic, heliocentric model.

At the same time, we are free to follow the *Bhāgavatam* and place the earth in the center. This leads to an interpretation in which the Fifth Canto offers a perspective on the solar system, rather than describing an absolute physical disk.

What do we see if we pursue this idea? It turns out that the combination of geocentric and heliocentric motion gives the planets spirograph-like orbital paths. These orbital paths are tangent to the circular features of Bhū-maṇḍala – a point that I discuss in detail in my book *Mysteries of the Sacred Universe*. Thus the circular features of Bhū-maṇḍala provide an accurate map of the solar system as seen from a geocentric perspective.

Our conclusion is that the *Bhāgavatam* can be understood from a multi-perspectival standpoint that emerges with clarity when the text is seen against a background of deep knowledge. This is a conclusion that could be of interest to scholars.

www.ingramcontent.com/pod-product-compliance
Lightning Source LLC
Chambersburg PA
CBHW050451110426
42744CB00013B/1959